Ernane Rosa Martins
Luís Borges Gouveia

Inverted Classroom using Mobile Learning

Ernane Rosa Martins
Luís Borges Gouveia

Inverted Classroom using Mobile Learning

ScienciaScripts

Cover image: www.ingimage.com

This book is a translation from the original published under ISBN 978-613-9-64756-9.

Publisher:
Sciencia Scripts
is a trademark of
Dodo Books Indian Ocean Ltd. and OmniScriptum S.R.L publishing group

120 High Road, East Finchley, London, N2 9ED, United Kingdom
Str. Armeneasca 28/1, office 1, Chisinau MD-2012, Republic of Moldova, Europe
Printed at: see last page
ISBN: 978-620-7-78690-9

CONTENTS

CHAPTER 1

INTRODUCTION

The consequences of the evolution of new technologies, centred on mass communication and the dissemination of knowledge, have not yet been fully felt in education. Education still coexists with written language, even though the current dominant culture is permeated by a new language, that of television, information technology, and especially the language of the *Internet.* The culture of paper is perhaps still the biggest obstacle to the intensive use of the *Internet.* For this reason, young people who have not yet fully internalised this culture adapt more easily than adults to using computers. They are already being born into this new culture, the digital culture (Gadotti, 2000).

Educational systems still work too much with traditional resources that don't appeal to children and young people. The computerisation of education aims to change teaching methods to improve thinking skills, rather than simply developing memory. The role of schools must increasingly be to teach critical thinking. To do this, we need to master more methodologies and languages, including electronic language. New technologies now make it possible to access knowledge transmitted not only in words, but also in images, sounds, photos and videos *(hypermedia),* and to create new spaces for knowledge. Now, in addition to the school, the company, the home and the social space have also become educational. *Cyberspace* has done away with the idea of time and space for learning; the place and time for learning is up to you, making students not just receivers of information, but also senders of information (Gadotti, 2000).

According to Silva, 2012, p.254, "teachers need to be aware of the specific movement of digital technologies in tune with the information society, *cyberculture and* the communication profile of learners", making it necessary to review the practices of education professionals, integrating new technologies into the teaching and learning process. There are many methodologies that make this process more dynamic by integrating digital technologies, including the *Flipped Classroom,* known in Brazil as the "Inverted Classroom", the name that will be used in this research.

The Flipped Classroom (FL) methodology proposes a reversal of traditional classroom practice, using digital technologies to help build knowledge through video lessons, games, audio files, *applets and* other tools. With the help of these resources, teachers can optimise classroom time and use it for interactive activities, deepening and discussing the subject (Barseghian, 2011).

The Flipped Classroom has been an alternative in school organisation, contributing to student independence in the construction of their knowledge. In this theory, the teacher's role becomes that of a mediator between knowledge and the student, promoting the active participation of students in the teaching and learning process (Schneider *et. al.* 2013).

In general, courses use lectures as their dominant methodology, developing students' skills of memorisation and reproduction. Given students' poor study habits, they generally don't develop learning autonomy and remain dependent on their teachers. Failures in the teaching and learning process can be due to the teaching methodology adopted by the teacher, the attitude of the student and some factor in the educational institution or a combination of the three causes (Frescki & Pigatto, 2009).

Reflecting on technology-orientated education, it is necessary to rethink educational parameters in order to modify the formulation of teaching activities associated with the use of computers or other digital media (Cabral, 2005). Bishop & Verleger (2013) explain the importance of student-centred learning theories, such as the Inverted Classroom (ILC) concept, which is an educational technique with interactive group learning activities in the classroom and individual guidance through digital technologies outside the classroom, making learning more dynamic through the integration of digital technologies. Its main characteristic is not to use classroom time for lectures.

With Digital Information and Communication Technologies (DICT), various forms of distance *learning* have emerged, including *blended learning,* which combines face-to-face and distance learning activities. The *flipped classroom* is one of these possibilities (Valente, 2014).

Blended learning mixes *online* and in-class moments, with interactions between students and teachers. During *online* activities, students control when, where, how and with whom they study. The content studied *online* must be prepared specifically for the subject. Face-to-face time allows for teacher supervision, values interpersonal interactions and complements online activities, providing efficient, interesting and personalised learning (Staker & Horn, 2013).

In SAI, content is studied *online* before classes, which now become the place to work on what has been studied, through practical activities such as problem-solving and projects, group discussions and laboratories (Valente, 2014). The Flipped Classroom is being adopted by several renowned international institutions, such as *MIT, Havard, Duke and Stanford* (Bishop, 2013). SAI can be an alternative in school organisation, contributing to student independence and the construction of knowledge. In this theory, the teacher becomes a mediator between knowledge and the student, promoting the active participation of students in the teaching and learning process (Schneider *et. al.,* 2013).

Our era is known as the age of knowledge, given the importance given to knowledge in all sectors today, especially as a result of computerisation and the globalisation of telecommunications. However, what we are seeing is the predominant dissemination of data and information rather than knowledge. This is possible thanks to new technologies that store knowledge, in a practical and accessible way, in huge volumes of information, allowing search and access in a very simple, user-friendly and flexible way. Not only access, but also as a sender of information. Through the *Internet,* from anywhere and at any time, you can access countless libraries in many parts of the world, allowing you to access knowledge not only through words, but also through images, sounds, photos, videos, etc. (Gadotti, 2000).

Communication technologies change some of the roles of teachers, but they don't replace them. The task of transmitting information is now done by databases, books, videos, programmes on CD. Teachers now have the role of stimulating the student's curiosity in wanting to know, research and seek out the most relevant information. It is the teacher who coordinates the process of presenting the results to the students, contextualises the results, adapts them to the students' reality and questions the data presented, as well as transforming information into knowledge and knowledge into knowledge, life and wisdom (Moran, 1995).

Technologies are bringing a new enchantment to the school, opening up its walls and enabling students to talk and research with other students from the same city, country or abroad, at their own pace. The same goes for teachers. Their research can be shared by other students and published instantly on the web. Numerous electronic libraries, online magazines, texts, images and sounds are available to students and teachers, making it easier to prepare lessons, do research and have attractive material to present. The teacher becomes closer to the student, being able to receive messages with doubts, pass on additional information to certain students, adapt their lesson to the pace of each student, seek help from other colleagues on problems that arise, new programmes for their area of knowledge. The teaching-learning process can gain more dynamism, innovation and communication power (Moran, 1995).

Today's world demands an increasingly critical and creative professional, with the ability to think, to learn how to learn, to work in groups and to realise their intellectual potential, with the capacity to constantly improve and refine ideas and actions. This new attitude cannot be transmitted but must be built and developed by each individual, in other words, it must be the result of an educational process in which students experience situations that allow them to build and develop these competences. In this context, computers and new technologies can be important allies in this process (Valente, 2012).

M-learning (mobile learning) refers to the learning process supported by the use of mobile and wireless information or communication technologies, where the main characteristic is the mobility of learners (Saccol *et. al.,* 2011). Mobile learning involves the use of mobile technologies, alone or combined with other information and communication technologies, making it possible to learn anytime and anywhere (Unesco, 2014).

Mobile technologies have recently become easier to access, but they still need to be more widely used in school activities. However, it is necessary for teachers, students and the school environment to adapt (Alencar *et al.,* 2015). Some of the positive aspects of using mobile technologies in the classroom are reported by Porvir (2013), such as: Extends the reach and equity of education; Optimises time in the classroom; Allows learning at any time and place; Builds new learning communities; Supports on-site learning; Brings formal and informal learning closer together; Provides immediate assessment and feedback; Facilitates personalised learning; Improves continuous learning; Improves communication; Maximises the cost-benefit ratio of education.

For Saccol (2011, p. 25), *m-learning:* refers to learning processes supported by the use of mobile and wireless information or communication technologies, the fundamental characteristic of which is the mobility of learners, who can be far away from each other and also in formal educational spaces, such as classrooms, training rooms, capacity building and training or even in the workplace.

The use of digital tools in the school context simply does not guarantee quality in education, it is necessary to transform teaching practices. Moreira & Kramer, (2007, p. 1038), state that: "it is as if technical objects could, by magic, guarantee quality in education. In many cases, the content traditionally taught in the classroom is transposed into new media".

Moran (2013, p.12) states that: "The school needs to relearn how to be an effectively meaningful, innovative, entrepreneurial organisation.
It is too predictable, too bureaucratic and not very stimulating for good teachers and students." The same author also states that: "While society changes and experiences more complex challenges, formal education is still generally organised in a predictable, repetitive, bureaucratic and unattractive way. Despite advanced theories, a conservative vision prevails in practice, repeating what is well-established, which offers no risk or major tensions.

Regarding the role of the teacher, Moran (2015, p. 24) mentions that they have the role of curator and advisor:

> A curator, who chooses what is relevant from so much information available and helps students find meaning in the mosaic of materials and activities available. Curator, also in the sense of carer: he takes care of everyone, gives support,

> welcomes, stimulates, values, guides and inspires. He guides the class, the groups and each student.

For Moran, the school has an obligation to make use of technology and get something out of it. It is unacceptable for teachers not to use technological resources in their lessons. Coscarelli and Ribeiro (2011, p. 91) collaborate by saying that:

> It's necessary to recognise the importance of technological discoveries, so that we don't stand still in time; but it's also necessary to rescue man's creative and productive sense, so that technology becomes an ally and not a mere object of domination or even alienation.

Moran (2013, p. 27) states that "a beautiful school also depends on an innovative project in which the *Internet* is included as an important methodological component". The use of Information and Communication Technologies in the classroom can bring significant improvements to learning, because, according to Moran (2013, p.31): "with current technologies, the school can be transformed into a set of spaces rich in meaningful learning, both face-to-face and digital, which motivate students to learn actively, to research all the time, to be proactive, to know how to take initiatives, to interact".

It is therefore essential that contemporary teachers seek to bring new digital technologies into their school teaching experience and practice, given that students are already using and adapting to these new technological trends.

Lévy (2010, p. 173) says that:

> The teacher's main function can no longer be the dissemination of knowledge, which is now done effectively by other means. Their remit must shift towards encouraging learning and thinking. The teacher becomes an animator of the collective intelligence of the groups in their charge. Their activity will be centred on monitoring and managing learning: encouraging the exchange of knowledge, relational and symbolic mediation, personalised politicisation of learning paths, etc.

According to Moran (2012, p. 7), "The school is unattractive, the subjects are loose, they talk about subjects with no direct connection to the student's life. Many teachers are demotivated." Therefore, the use of digital tools, when done in a planned way, can contribute to the development of students' skills and competences in an autonomous way based on their needs, facilitating and bringing more innovation and dynamism to their studies. Schools and teachers need to update their pedagogical practices to promote the use of new digital tools in academic activities, in order to keep up with students' progress.

Gomez (2010, p. 88-99) states that: "Social networking programmes, whether personal, thematic or professional, were not actually created for educational activities."The same author also states that the pedagogical use of online social networks *and* applications for mobile devices requires a certain amount of care, saying that: "The web is yet another space in the contemporary school that needs guidance and care if it is to be transformed into a pedagogical device."

CHAPTER 2

FLIPPED CLASSROOM

2.1. Flipped Classroom: Characterisation

The flipped classroom, according to its creators Jonathan Bergmann and Aaron Sams, is the concept in which what used to be done in the classroom in the traditional model is now carried out at home, while activities that used to be carried out alone by students as homework are now carried out in the classroom (Bergmann & Sams, 2016). This is an approach whereby the student takes responsibility for their theoretical study and the classroom lesson simply serves as a practical application of the concepts previously studied (Jaime; Koller & Graeml, 2015). In other words, the teachers make the content available *online* to the students, who study it beforehand, and during the class they carry out practical activities such as problem-solving, projects, group discussions and laboratories (Valente, 2014).

In this methodology, the student has the first contact with the content virtually, outside of school, and then discusses and answers questions during the lesson. In the inverted classroom there is a paradigm shift. The premise is to consider the knowledge that the student is capable of constructing autonomously when provided with the conditions by the teacher. The consolidation of learning takes place in the classroom, where questions are clarified and issues are problematised in discussions fostered by the content provided by the teacher or made available by the students themselves (Souza, 2015). In this context, the teacher has a great responsibility in choosing the type of technological resource they will use to support their teaching programme, dedicating themselves to planning relevant practices so that the technology in use is a bridge to building meaningful learning (Souza, 2015).

According to Valente (2014, p. 86), after 2010, the term "Flipped *Classroovt*" began to be used in primary and secondary schools, mainly due to numerous international publications. However, the first studies are not recent and date back to the 1990s (Trevelin, Pereira & Oliveira Neto, 2013; Teixeira, 2013 and Valente, 2014). In the United States there is an organisation, the *Flipped Learning Network* (FLN), which disseminates concepts about flipped learning so that educators can implement it successfully. This theory has been tested and approved by several US universities, such as *Duke, Stanford, Harvard* and *Massachutsetts Institute of Technology - MIT,* and in American K-12 education. The flipped classroom theory has become a growing trend in education in several countries, such as Finland, Singapore, the Netherlands and Canada (Ramal, 2015). In Brazil, there are several schools and universities that apply this approach, such as: Colégio Dante Alighlieri, UNIAMÉRICA, UNISAL, PUC do Paraná and Universidade Positivo.

The theory of the inverted classroom is a model based on hybrid teaching. Hybrid teaching "is a combination of the resources and methods used face-to-face and *online, which* seeks to capitalise on the advantages of either learning system" Miranda (2005, p.48). Hybrid teaching can also be called *blended learning* or *b-learning,* which arose from *e-learning* experiences (Tarnopolsky, 2012, p.14).

According to Lima & Capitão, 2003, p. 38, *e-learning* involves "web-based learning", "*Internet-based learning",* "online learning", "distributed learning" and "computer-based learning". Christensen, Horn & Staker, 2013, p. 33, say that the Flipped Classroom concept is a technique in which teachers can improve student engagement. It can be simpler to implement, requiring only good planning on the part of the teachers.

According to Bergmann & Sams (2016) the SAI model consists of: "what is traditionally done in the classroom is now done at home, and what is traditionally done as homework is now done in the classroom". According to the same authors, in this model time must be restructured. At the beginning of the lesson, it should be used to answer questions about the video watched at home, then guided and independent practicals and/or laboratory activities should be carried out. In this way, students apply the theoretical content to classroom practice, in the form of problem-solving. The teacher acts as a mediator while the students actively participate in the teaching-learning process.

Bergamann & Sams (2016) state that "the teacher's role in the classroom is to support the students, not to transmit information". Another important factor in this model is the use of ICT, which makes it possible to make better use of classroom time and encourages student participation. According to Bergamann & Sams (2016), when content is transmitted *online, it* allows students to organise their own pace of study and learning in a flexible way.

According to Moran (2014), the flipped classroom is one of the most interesting theories today, as it mixes teaching methodology with technology, concentrating basic content in the virtual environment and creative and supervised activities in the classroom, such as challenge-based learning, projects, real problems and games. As there is no single model of classroom inversion, the teacher is free to carry out practical activities as he or she wishes, which allows students to work in groups or on individual activities (Bergmann & Sams, 2016). Moran (2014) explains that the role of the teacher is that of curator and advisor, "Curator, in the sense of carer: he is attentive *to* each one, gives support, welcomes, stimulates, values and inspires. He or she guides the class, the groups and each student."

2.2. How does the flipped classroom work?

In the flipped classroom theory, students have access to the content before the lesson and the first few minutes are used to clarify any doubts, in order to clear up any misunderstandings before the concepts are applied in the practical activities, which require more time (Bergmann & Sams, 2016).

Within the classroom, activities take on various cognitive forms, such as: applying, analysing, evaluating, creating, relying on the support of their peers and teachers. This makes it possible to prepare learning activities in advance, which helps students develop their communication and thinking skills (Lage; Platt & Treglia, 2000).

Moran (2014) explains that the role of the teacher is that of curator and advisor, "Curator, who chooses what is relevant from so much information available and helps students find what they are looking for. Curator, also in the sense of carer: they are attentive to each individual, supportive, welcoming, stimulating, valuing and inspiring. They guide the class, the groups and each student. They have to be competent intellectually, affectively and managerially (delivering multiple and complex learning)." Figure 1 summarises the Flipped Classroom methodology.

Figure 1. Summary of the Flipped Classroom methodology. Source: Barbosa, et. al. 2015.

2.3. Tools used in the flipped classroom

Although the Flipped Classroom theory is not a new teaching methodology, it has recently become more prominent due to the evolution of digital technologies, which have made it possible to use a vast amount of resources to plan and implement this model, thus promoting the integration of digital technologies into learning (Teixeira, 2013).

Here are some of the types of tools that can be used with the Flipped Classroom: Video lessons; *Podcast, vodcast* and *screencast;* Digital whiteboards; Teaching platforms; Forums, *wikis,* self-assessment and peer-assessment questionnaires, present in Virtual Learning Environments (VLE) such as Moodle, Edmodo and others; Courses and learning objects from repositories such as *Coursera* and *Khan Academy;* Social networks.

Bergmann & Sams (2016) use videos in place of direct instruction, but state that the flipped classroom is not synonymous with *online* videos

or replacing teachers with videos, because it is precisely the face-to-face interactions and activities that are most important in this methodology. There are many cases of teachers applying the inversion concepts without using videos.

2.4. Advantages and Disadvantages of the Flipped Classroom Model

Bergman & Sams (2016) point out some of the advantages of SAI: flexibility of time, being able to access the activities *online* at the time and place they want; allows the teacher to dedicate more time in the classroom attending to students with doubts; allows students to pause the videos and understand the concepts in their own time; intensifies the relationship between teacher and student; increases interaction between students, with group activities; it changes classroom management, getting students more involved in activities; it makes the classroom more transparent; students progress at their own pace, being able to study whenever they want and take as long as they need to finalise, view and revise the online materials as many times as necessary*; the* teacher can reuse the lessons the following year.

Moran & Milsom (2015) observed better student performance in the assessments, the class was more involved in the development of the content and the students felt more confident to learn independently. Tune; Sturek & Basile (2013) also noted that by carrying out the activities before class there was an improvement in the discussions held in class and that the students did better in the assessments than in the traditional course.

Pavanelo & Lima (2017) point out some of the advantages of SAI: it requires changes in the teacher's posture, it needs the choice, elaboration of efficient teaching materials; changes in the students' posture; existence of technical problems such as not having access to the *internet and* thus not carrying out the tasks; need for a lot of organisation, needing to familiarise students with a new and different concept; need to make the transition from a passive learning model to an active one; need to motivate students to do their tasks and prepare for classes, otherwise they will be easily distracted by other things; requires a lot of self-discipline, they need to know how to study, which takes time.

Valente (2014) identified some positive points regarding the use of this methodology, such as: the possibility for students to work at their own pace and develop as much understanding as possible; prior identification by students of points that need to be better assimilated and the formulation of doubts that can be clarified in the classroom; the possibility for teachers to customise classroom activities according to student needs; encouraging social exchanges between colleagues through classroom activities. The same author also identified some concerns and criticisms about the methodology, such as: teachers' worries about the difficulties that students may have, due to the way this methodology is proposed; the dependence on technology for its realisation, which can create an unequal learning environment and the possibility of students not being prepared before the class and not being able to carry out the face-to-face activities.

Bergmann & Sams (2016) present some of the advantages of using the inverted classroom, such as: each student can be assisted at their own pace, as many times as they need, and they can ask their parents or colleagues for help if necessary. In the classroom, the teacher can guide activities according to the specific needs of each student.

2.5. Comparing the Inverted and Traditional Classroom Models

Basically, an example of a comparison between the traditional and flipped classroom models was the work by Trevelin; Pereira & Oliveira Neto (2013), entitled "The Use of the 'Flipped Classroom' in Higher Technology Courses: Comparison Between the Traditional Model and the Flipped Classroom Model Adapted to Learning Styles". This work aims to compare the pass and fail results of some classes taking the subject Operating Systems at the Taquaritinga College of Technology, using the two methodologies. The sample consisted of 148 students from the second semester of the Systems Analysis and Development course, who were divided into four classes. Three classes were taught in the traditional way. In the fourth class, the lessons followed the Inverted Classroom methodology. At the end of the semester, there was an improvement in the failure rate, which fell considerably, and around 90 per cent of the students said in a questionnaire that they preferred the flipped classroom methodology to the traditional one. Table 1 shows a comparison of the performance of the two models.

Table 1. Comparison between the inverted and traditional classroom models

Period	Traditional model	SAI model
Before school	There is no prior contact with the content that will be worked on in class;	Students have prior contact with the content they will be working on in class;
During lessons	The teacher presents the content in an expository manner, using the blackboard as a teaching resource;	The teacher answers questions; development of guided practical activities; carrying out case studies in groups or laboratory activities.
After school	Doing work or exercises at home.	Application of the concept in a real situation.

Table 2 was inspired by Gannod et. *al.* (2008) and shows the main differences and similarities between the two models.

Table 2. Comparison between the traditional and inverted models, Gannod et. al. (2008)

Factor	Traditional model	SAI model
Lesson preparation time by the teacher	Study of the content, preparation of activities to be carried out in class;	Preparation of asynchronous materials (videos, comprehension exercises, etc.); preparation of activities to be carried out in class;
Role of the apprentice	Listening to explanations in class; carrying out activities in class and at home, in groups or individually;	Study beforehand at home; carry out activities in class, preferably in groups;
The teacher's main role	Content transmitter;	Mediator and facilitator;
Activities carried out in the classroom	Explanation of content; resolution of doubts; exercises to fix the content explained in class;	Varied activities to fix and deepen the content seen at home; resolving doubts;
Activities carried out at home	Varied activities to fix the content seen in class;	Access to asynchronous materials and simple activities to understand them, such as *quizzes;*
Where to access the content	Primarily in the classroom;	Primarily through asynchronous materials that can be accessed from different locations;

It's important to note that students who are unable to attend one or more lessons for any reason may miss out on a significant part of the content taught in class, but in

SAI the content is always available to the student (Bergmann & Sams, 2012).

2.6. Using Flipped Classroom Theory in Teaching: A Systematic Literature Review

With the possible publications arising from the pedagogical possibility of using the Flipped Classroom in mind, we decided to systematise a study in order to map and find out what researchers have discovered when they use it with their students.

A systematic review summarises the evidence related to a detailed intervention strategy by applying a research protocol. This protocol is systematised and has its own characteristics for searching, analysing and synthesising the data obtained. To develop this protocol, it is necessary to define a series of criteria that underpin the entire research. All these criteria must be impartial and cannot be altered once the protocol has been implemented (Kitchenham, 2004). The review protocol makes this type of research different from the narrative review, which is based on an opinionated nature, based on the interpretation made by the researchers. This means that this review has less scientific evidence than a systematic review.

In order to develop a systematic review, it is necessary to make sure that all the important articles or those that could have an impact on the conclusion of the review are included, so it is necessary to establish an effective strategy to include the most appropriate terms in the search and also a good choice of specific databases on the subject (Sampaio & Mancini, 2007).

In order to select the sample used, some inclusion and exclusion parameters were defined. The aim of this procedure was to help compile a set of articles on the same subject. The factors used in this study are shown in Table 3.

Table 3. Inclusion and exclusion factors.

Inclusion Factors	Exclusion factors
Articles written in Portuguese; Articles available on *Google, Google* Scholar, Periódicos Capes and *Scielo;* Studies on Flipped Classroom Theory; Empirical or theoretical articles.	Articles in a foreign language; Articles in databases other than *Google, Google* Scholar, Periódicos Capes and *Scielo.*

To compose the documentary database of articles that formed part of this study's sample, a systematic web search was carried out using the *Google, Google*

Scholar, Periódicos Capes and *Scielo search* engines. To search for the articles in the databases, the following words were used as descriptors: "Inverted Classroom Theory", "Inverted Classroom", "Inverted Classroom Theory and Education", "Inverted Classroom Theory and Teaching", "Inverted Classroom and Teaching" and "Inverted Classroom and Education".

In the search for articles to carry out the research, after excluding duplicate articles that met the established criteria, a total of 62 articles were found. These were then exported to the *ATLAS.ti* software and analysed, coded and categorised.

The results obtained answer the questions: What are the reasons for implementing the Flipped Classroom? What are the roles of students and teachers in the Flipped Classroom? What are the benefits and challenges of the Flipped Classroom? What are the main types of studies carried out? In which areas and contexts is it most widely used? What remains to be explored? What technologies are being used in the Flipped Classroom? What level of education is the Flipped Classroom being used in? What type of teaching is the Flipped Classroom being used in? What is the acceptance of this type of methodology? Have students evolved with this type of methodology?

By coding the articles in the *ATLAS.ti* software, it was possible to find the possible reasons for implementing the flipped classroom, as shown in Table 4.

Table 4: Reasons for implementing the IAS.

Reasons for implementing SAI
Attend to students who miss classes for any reason Remodel and propose changes to the teaching-learning process Facilitate the understanding of concepts that are not usually easy for students to grasp
Changing the way content is passed on to students, avoiding a simply mechanical approach
Facilitating the understanding of concepts that are not easy for students to grasp Greater use of class time
Encouraging students to work actively in class, making them more involved in the lesson, rather than just listeners.
The high failure rate, which leads to students being held back Technological advances, as the *internet* and other technologies now provide easy access to much of the information transmitted in the traditional classroom
Problems with some students' lack of financial resources to buy new, up-to-date textbooks
Improved student development, resulting in more interested and engaged students who seek success in their academic and professional careers. A more student-centred approach

The articles analysed also revealed the new role that the student and teacher have in the Flipped Classroom approach, shown in Tables 5 and 6 respectively.

Table 5. Role of the student in the SAI approach.

The student's role
Watch the video lessons or materials made available before the class Participate in classroom activities Be responsible for your own learning Complete all activities on time Demonstrate what you know and investigate what you don't understand Study at your own pace

Table 6. Role of the teacher in the SAI approach.

The teacher's role
Mentor Facilitator
Helper Provide individualised support to all students Encouraging Inspiring Questioner Listening to ideas

In the articles used, it was possible to find many benefits and challenges to be faced when implementing the Flipped Classroom proposal. Tables 7 and 8 show the benefits and challenges of the Flipped Classroom approach respectively.

Table 7. Benefits provided by the SAI methodology.

Benefits of SAI
Replacing passive learning with participatory lessons The teacher can produce the content only once on video, and can repeat it in several classes, as long as the information is kept up to date Permanent availability of content for students Allows you to work on content in a broader and deeper way Explaining all the planned content in less time than in traditional classes Freeing up class time for more engaging activities (laboratory research, problem solving and projects) By recording the video, the teacher reflects on the lesson Increased student-student interaction Increased student-teacher interaction Provide individualised support to students Improve relationships Student progress in assessments, commitment and attitudes Facilitating the identification of student difficulties Parental involvement, monitoring their children's education at school Increasing student responsibility Students work at their own pace and in their own style Promoting the development of communication skills Promotes teamwork Promotes collaboration of ideas Allows students to put their learning into practice

Table 8. Challenges of the SAI methodology.

SAI's challenges
More time to prepare lessons

Students need to be prepared and orientated for the development of the lesson, as it is a different methodology to the traditional one Student resistance to the new method Not all areas allow flipping classes, and it is more likely to be used in more didactic courses Technology-dependent model, some students may not have access to information at home
The lack of computers or *internet* in students' homes Demand that the approach can make on home computers, if there is more than one student in the same house Students don't access videos or other media at home Students don't study before class and therefore can't follow the development of the classroom lesson. Little participation from students in discussions, they just observe The conditions in which students come into contact with the content, not at the same time as other activities

In the articles used, it was possible to see which were the main types of studies carried out using the Flipped Classroom approach. As illustrated in Table 9.

Table 9. Main types of studies carried out with SAI.

Main types of studies carried out with SAI
Address the use of *Internet-based* collaborative tools to provide interactivity in the school environment To analyse the possibilities and challenges of the SAI, in order to demonstrate its potential for teaching and learning processes in higher education To point out the potential, some of the problems faced and the opinion of students in relation to the SAI methodology To present the importance that SAI brings, combined with the technology of the *Facebook* virtual social network and teacher mediation, with significant gains in carrying out pedagogical work, characterising it as a hybrid model that does not displace the traditional classroom model To present the difficulties encountered, the need for student interaction and inverted preparation, as well as the knowledge acquired through the use of video lessons and other interactive resources such as texts, games, audio files, video conferences, forums, etc. Present the difficulties faced by teachers working in higher education Present experiences of hybrid teaching, through the application of the SAI methodology Present the results of experiences using the SAI method, reporting on students' perceptions of this teaching strategy Present reflections and reports as a result of the use of two of the active methodologies in university teaching, represented by the SAI and PBL - Problem-Based Learning methods, highlighting the pros and cons pointed out by higher education teachers Present varied technological support, based on free web 2.0 tools, which provide opportunities to incorporate creative learning strategies within this new teaching and learning model, with a view to maximising the construction of student knowledge To present a brief discussion of the way in which reading teaching can be developed through active learning methodologies, in this case, in particular, through the proposal of the inverted classroom in conjunction with the *GoConqr* teaching platform. Evaluate the application of SAI in higher education Search for innovative methods for the Virtual Teaching and Learning Environment of a Higher Education Institution Characterise the evolution of the concept of flipped learning, identifying the structuring pillars of this learning and presenting a reference for its use in distance education.
distance Characterise and reflect on SAI Classify the assumptions of active methodologies, with a focus on Problem-Based Learning, the Inverted Classroom and Peer *Instruction,* discussing possible adaptations to the national context, taking into account the peculiarities of our schools and the academic community. To compare the results of a course taught to different classes, either in the traditional way or by combining learning styles with the SAI methodology To prove that by implementing the SAI

model, the students' teaching-learning process was better than the traditional process.
To conceptualise hybrid teaching, raising some questions about the role of the teacher and the student in this new methodological perspective. Contextualise hybrid teaching within the critical social pedagogical trend of content with a focus on higher education. Present the concept of the inverted classroom and the role of the teacher in this change of educational paradigm.
To see that digital devices linked to the *Internet* enable new ways of teaching and learning and allow for the articulation of new teaching methodologies.
Contribute to improving student learning by making lessons attractive, objective and student-centred.
Demonstrate the efficiency and innovation of SAI as a collaborative teaching method between teacher and student
Demonstrate the potential of the SAI methodology by designing and evaluating a process that supports collaborative learning in its implementation in the classroom.
Describe the essential aspects of the "inverted classroom" in order to help those who are interested in using it.
Describe a Systematic Literature Review to answer some questions about the SAI methodology
Discuss the SAI teaching strategy as an active methodology and present the students' perceptions
Discussing the relationship between mobile digital technologies and education through an SAI experience
Discuss the different modalities of *blended learning* and the flipped classroom, how ICT is used in different models of implementing this pedagogical approach, how the flipped classroom can be implemented and the positive and negative points about the flipped classroom presented by different authors.
Discuss a new proposal for a teaching-learning methodology that will be used in an educational institution Identify challenges and advances in the use of SAI
Encouraging active learning and analysing class behaviour with the SAI Investigating how the flipped classroom, which has already been successfully applied in the face-to-face modality, could be used appropriately in distance learning courses
Investigating and reflecting on the possible impacts of implementing the Flipped Classroom, a hybrid teaching method
To investigate, based on empirical research in databases, teachers' perceptions of the possibilities and challenges offered by the flipped classroom proposal.
Survey of publications involving SAI, highlighting its relevance and application Propose solutions to be adopted in the teaching and learning process
To provoke critical reflection on the Inverted Classroom as a methodological option with a view to teaching quality
Carry out a literature review and critical analysis of the concepts of SAI

Report on the experience of using the *WhatsApp* social network in the teaching and learning process of the proposed content
Report on the implementation of the SAI pedagogical model in higher education distance technology courses, using face-to-face and distance meetings, better known as the *Blended Distance* Learning methodology.

It was also possible to verify the areas and contexts where the Flipped Classroom approach is most used. As shown in Table 10.

Table 10. Areas and contexts where the SAI approach is most used.

Areas and contexts where SAI is most used
Various experiments carried out by educators from different areas and educational levels, mainly in secondary education and in higher education courses in different areas
Hybrid courses that seek to take advantage of each teaching modality, with content offered at a distance and in person, generally do not give up traditional teaching
Various areas of human activity, such as the academic environment, whether through the creation of new educational models or the development of digital tools to support teaching and learning processes.
Advances in technology, mainly focussing on communications and access to information, enabling a society geared towards learning without place, time and source of specific knowledge
The need for continuing training for teachers to not only understand the logic of the inverted classroom, but also to reflect on the role of higher education today and its function
Teachers haven't got to grips with digital media in order to review their pract ce and understand a change that isn't one-off, but happens all the time, redefining roles and new ways of thinking
Students are already being born into the digital culture, which is something familiar to them, and the school needs to use technological tools in its favour, contributing to quality education
Mediated by interactive technologies such as virtual environments and collaborative tools where students work in groups, face-to-face or remotely, guided by teachers, the classroom is transformed into a collaborative, open and continuous learning space
Difficulties for educational agents to change. Teachers need a radical change of attitude in the classroom, both in the preparation of teaching materials and in the methodo ogy and pedagogical strategy in the classroom, bearing in mind that they have not been trained in the logic of the flipped classroom. Students need to be committed and take responsibility for their learning

We also looked at what remains to be explored in the Flipped Classroom approach. As illustrated in Table 11.

Table 11. What remains to be explored in the Flipped Classroom approach.

What remains to be explored at SAI
To analyse the causes of lack of time and difficulties in devoting oneself to the subjects, such as: working during the day or lack of *internet* To check the effectiveness of *online* tutoring
Investigate the possibilities of training and education for teachers, presenting new practices and teaching resources
To study the reasons for resistance to change, because it is a different proposal, because it requires them to work at home instead of being exposed to the content for the first time in class To propose a reformulation of the teacher and pedagogue training curriculum for new forms of learning using active methodologies To investigate new models of teacher performance in the classroom To verify how and for what purpose students use technologies in class, To investigate how teachers are exploiting technological resources and integrating them into school activities.
Although there are currently many discussions about the Flipped Classroom. Bishop and Verleger (2013) report a certain lack of consensus on its exact definition, due to a limited amount of academic research into its effectiveness. Investigating ways to dynamise and improve lessons.
Contribute to the discussion of the difficulties in maintaining academic standards in larger and more diverse classes
Verify the concerns of higher education teachers, such as: the lack of institutional support for the use of active methodologies
Observe factors that motivate students, such as topics of interest, freedom to express ideas or

relate content Investigate how managers view SAI

As the analysis of the articles showed, various technologies are being used in the Inverted Classroom proposal. As shown in Table 12.

Table 12. Technologies used in the Flipped Classroom approach.

Technologies used in SAI
High-quality visual and educational videos Digital and interactive games Virtual learning environment Virtual libraries *Internet* use Use of mobile technologies such as *tablets* and *smartphones* Virtual social networks Book or television Discussion forums and *quizzes*

The articles show that the Flipped Classroom approach is being used at some school levels. As Table 13 illustrates.

Table 13. Level of school where the Flipped Classroom approach is used.

Level of schooling where SAI is used
Higher Education High School Lato sensu and stricto sensu postgraduate courses

The articles also show that the Flipped Classroom approach is being used in some teaching modalities. As shown in Table 14.

Table 14. Teaching methods in which the Flipped Classroom approach is used.

Teaching methods in which SAI is used
At a distance Face-to-face *Blended learning* (face-to-face and distance learning activities)

We can also see the difficulties in accepting the Flipped Classroom approach. As Table 15 shows, there are many difficulties in accepting this new teaching method.

Table 15. Acceptance of the Flipped Classroom approach.

Acceptance of SAI
School management - the teacher usually needs the acceptance of the school management to use new methodologies, and some schools are resistant to change, opposing any less conservative educational initiative Teachers - some teachers don't feel very excited about this methodology Students - at first the students have been very receptive to it, as it is an innovative proposal that presents a new way of perceiving and organising the school environment, but like many other innovative methodologies, it has also received some negative criticism, such as: dependence on technology

Finally, the students' progress with the use of the Flipped Classroom approach was checked. As illustrated in Table 16.

Table 16. Students' progress with the Flipped Classroom approach.

Students' progress with SAI
Mixing new teaching technologies with interactive practical lessons resulted in improved learning

The effectiveness of the methodology depended on the social, political and cultural context in which the school was located
The majority of students considered the proposal to be valid, highlighting aspects such as more dynamic and personalised lessons, making the most of time and the use of digital technologies
Significant increase in student performance in assessments, high participation and good acceptance
More commitment to practical group activities during lessons
Increased class attendance, involvement, encouragement, confidence, interest, reading, group discussions, autonomy in learning and understanding of content

Various improvements to education involving new technologies are currently being studied. And schools need to learn how to apply these studies and new technologies in their favour, since students are already being born into this digital culture. This research deals with the Inverted Classroom methodology, which uses technology in the development of the lesson, providing many benefits and challenges and knowing them, as well as their characteristics, is extremely important for their use.

One of the main points of the Flipped Classroom methodology is that students need to be prepared and orientated for the development of the lesson, they must be aware of what they need to understand in the tasks, work collaboratively, solve the problems assigned and be disciplined.

The research made it possible to find out: the reasons for implementation, the roles of students and teachers, the benefits and challenges, the main types of studies carried out, the areas and contexts in which it is most widely used, what still needs to be explored, the technologies, the level of education and the type of teaching in which it is being used, the acceptance and evolution of students with this type of methodology.

CHAPTER 3

M-LEARNING: MOBILE LEARNING

3.1. *M-Learning:* Overview and Definition

The popularisation of the *internet and* its media has changed Brazil's educational landscape through the use of virtual and digital media and mobile devices at school, meaning that these have become integrated as pedagogical tools for acquiring knowledge. As a result, teachers are faced with the challenge of keeping up with this pace, seeking to integrate students' out-of-school culture (Sena & Burgos, 2010). Wains & Mahmood, 2008 define *M-learning (mobile learning*) as "an emerging field that encompasses wireless technologies and mobile computing to enable learning to take place anytime and anywhere, maximising students' freedom". *Mobile Learning* or *m-Learning* arose from the availability of mobile devices and considering the specific needs of education and training (Nyiri, 2002).

Technological innovations stemming from the development of telecommunications are providing access to different environments and forms of learning. Previously, a device connected to a fixed network structure was needed, but now mobile devices allow access to similar educational environments and resources (Mulbert & Pereira, 2011). The mobile phone is the most popular and accessible of the devices that can support *Mobile Learning, and* does not require financial investment on the part of institutions, as it is a common tool and available in students' daily lives "If the computer is still a restricted object, the mobile phone is present in most schools,

in the backpacks of students from different social classes" (Merije, 2012, p.81). *Smartphones* are a technology that brings together various media in a single device (phone, *internet,* games console, personal computer resources, etc.). Combining the resources of mobile phones and mobile phone networks with those of the *internet has* made it possible to access and share content, which has given a new dynamic to communication processes, as well as learning (Merije, 2012). The digital revolution, the miniaturisation of devices and connectivity to communication networks have made it possible to mix the digital with the physical, creating an environment of semantic and cognitive technology that has reshaped the ways we do, create, think and relate to everyday life, work, home, leisure, education or any other space (Cordeiro & Bonilla, 2015).

According to Marçal *et al,* 2005, has emerged as an important alternative for distance learning and training, with the following objectives: Improve the student's learning resources, as they can rely on a computer device to carry out tasks, write down ideas, consult information via the *Internet,* record facts using a digital camera, record sounds and other existing functionalities; Provide access to teaching content anywhere and at any time, depending on the device's connectivity; Increasing the possibilities of access to content, increasing and encouraging the use of services provided by the institution, whether educational or business; Expanding the teaching staff and learning strategies available, through new technologies that support both formal and informal learning; Providing the means to develop innovative teaching and training methods, using new computing and mobility resources.

According to the 2014 *United Nation Educational, Scientific and Cultural Organisation* (UNESCO) document entitled "The Future of Mobile Learning", mobile learning through distance education (DE) is becoming an area of focus, as the use of mobile devices in formal education systems is increasingly becoming a popular model for study and research around the world. There are at least two main models for using mobile technology: One Computer per Student (UCA), in which each student uses a computer for school use, and *Bring Your Own Device (BYOD),* in which each student uses their own mobile device. The *BYOD* model has caused unprecedented changes in higher education and distance learning, allowing more and more students to access different teaching materials via mobile technology. Taking advantage of this increase in the number of people who have access to or own a mobile device, *BYOD* initiatives look promising for students all over the world (Unesco, 2014).

In higher education courses, the use of mobile devices such as *smartphones* and *tablets is helping students to* read texts on specific apps, replacing paper with a digital screen and reducing the number of books and notebooks. They also make chats available via communication apps such as *WhatsApp* and *Messenger,* enabling individual or group instant *chats* between teachers and students, helping to minimise difficulties. Another important issue is the possibility of offering a digital space, where teachers and students can access it whenever they want and at the most convenient time, being able to read texts, listen to and/or watch class resources, communicate instantly or publish their work at any time, as long as there is an *internet* connection (Boll; Lopes & Luchini, 2016). The importance of inserting new tools into the school context is notorious, but mobility, an essential characteristic of mobile devices, allows them to be used in environments outside the classroom too. This makes it clear that we need to encourage the learning process to go beyond traditional parameters, enabling the appropriation of knowledge in out-of-school environments too (Santana, 2016).

Access to digital technologies can enable the decentralisation of knowledge, diversity and the promotion of collaboration between cultural producers and social transformation (Knobel & Lankshear, 2007). Communicational mobility transcends time and space; access to knowledge is increasingly related to the connective space of the communicative network, depending on the mobility of digital culture. Thus, the concept of mobile learning is related to expanding the possibilities of the communication process between connected students and teachers (Lemos & Josgrilberg, 2009). Mobile technologies have become part of children's lives all over the world, so many governments and schools are using these devices for a range of different teaching and learning purposes, and promoting various studies in this area, showing that they help promote the learning, skills and perspectives that children need to compete and co-operate today. Focussing on the development of skills such as collaboration, critical thinking and problem-solving. Some projects explore the innovative features of mobile devices, others rely solely on *standard* features, others on personalisation capabilities, and others look at how mobile devices can encourage collaboration in teamwork (Moura, 2009).

There are countless new pedagogical challenges presented by the use and integration of ICT, forcing us to redefine the roles of the different partners in the educational process. Thus, ICT can be presented as a reinforcement of traditional teaching methods or as a way of renewing learning opportunities (Barros, 2011). Certain requirements are indispensable for technologies to become attractive to users. These requirements, according to Lee et *al.* (2004), are: Mobility, the ability to move the device at any time and in any place; Portability, the ease with which devices can be transported, and they must be geometrically comfortable and lightweight; Usability, the functional usefulness of the device, which must be usable and flexible for any type of individual, especially those with special needs; Functionality, offering numerous options for use, in order to meet the demands of users. According to the same authors, these requirements attract young people and adults from more recent generations as the main target audience. Users with this profile find in these new technologies *(Short Message Service (*SMS), *Facebook, Skype, WhatsApp,* etc.) the possibility of entertainment and education together through textual communication.

Technologies do not transform educational contexts. Simply inserting them into education, without creating a consistent, coherent didactic-pedagogical proposal and planning aligned with the needs of the students, does not allow us to exploit their potential and enable the development of differentiated educational practices. There are many possibilities, but technologies will only be used well in the educational context if the methodological proposals are sufficiently open, creative and learner-focused (Costa; Silva; Cordeiro & Silva, 2014, p. 59). According to Lucena (2016, p. 287), technologies applied in the educational sphere act as pedagogical tools that "enhance the production of knowledge built collectively and collaboratively, using socio-digital networks".

3.2. Mobile Technologies in the Educational Context: A Systematic Literature Review

With the possible publications arising from the pedagogical possibility of using mobile technologies in mind, it was decided to systematise a study in order to map and find out what researchers have discovered when using these technologies with their students.

Nowadays, a large number of students and teachers own mobile phones, almost all of which are *smartphones* with *internet* access. We also realise that they have great potential and possibilities for use in teaching all areas of knowledge. This can provide rich experiences for teachers and students, as well as greater sharing of information.

The procedure used to carry out this research was a systematic literature review, which is an effective methodology for integrating information from a set of studies carried out separately, i.e. investigating the current state of the art. These studies can present conflicting and/or coinciding results, as well as identifying new themes that need to be researched in accordance with scientific precepts, in order to gain a better understanding, helping to guide possible future research (Sampaio & Mancini, 2007). For Gonçalves, Nascimento & Nascimento (2015, p. 194), a systematic literature review "enables an investigation that aims to identify evidence related to a specific research problem, with the aim of highlighting ideas, positions and opinions of authors published in the area of knowledge in which it is inserted".

According to Gonçalves, Nascimento & Nascimento (2015, p.195), the main stages for preparing a systematic review are: Research Problem (question to be investigated); Research Protocol (careful description of the study); Databases (location of studies);

Inclusion/Exclusion Criteria (characteristics and specificities of the studies); Analysis, Critique and Evaluation (validity of the selected studies); Preparation of the Summary (Synthesis of the content covered); Identification of Evidence (studies grouped according to similarity); Conclusion (scope of the evidence identified). In order to select the sample used, certain inclusion and exclusion parameters were defined. The aim of this procedure was to help compile a set of articles on the same subject. The factors used in this study can be seen in Table 17.

Table 17. Inclusion and exclusion factors for the studies found.

Inclusion Factors	Exclusion factors
Articles available on *Google, Google* Scholar, Periódicos Capes and *Scielo;* Studies on Mobile Technologies in an educational context; Empirical or theoretical articles. Articles that address the teaching-learning process through mobile learning	Articles in databases other than *Google, Google* Scholar, Periódicos Capes and *Scielo;* Studies on Mobile Technologies in contexts other than education. Articles with no focus on teaching or learning

To compile the documentary database of articles that made up the sample for this study, a systematic web search was carried out using the *Google, Google* Scholar,

Periódicos Capes and *Scielo search* engines. To search for articles in the databases, the following words were used as descriptors: "Mobile Technologies", "Mobile Technologies and Education", "Mobile Technologies and Teaching" and "Mobile Technologies and the Classroom". In the search for articles to carry out the research, after excluding duplicate articles that met the established criteria, a total of 174 articles were found. These were then exported to the *ATLAS.ti* software and analysed, coded and categorised. By coding the articles in the *ATLAS.ti* software, it was possible to find the possible benefits of using mobile technologies in education, as illustrated in Table 18.

Table 18. Presents the benefits of using mobile technologies in education.

Benefits of using mobile technologies in education
Abandon the mass system and adopt an interactive system
Encourages students to make discoveries on their own
Introduces students to a research process
Develops critical thinking rather than memorising the information transmitted
Encourages students to collaborate with each other
Flexibility in when to study, where and for how long
Transfers tasks beyond the physical environment of the classroom
Broadens learning experiences inside and outside the classroom
Replaces the need for a computer to access the *internet* in the classroom
Facilitates communication and information exchange at any time and place
Allows socialisation and inclusion of people with disabilities or restricted mobility
Enables the use of video and audio resources
Provides a style of education focussed on individual learning demands and needs

For Feenberg (2010), the concept of education centred on human interactivity facilitates the participation of less favoured groups and can increase the cultural level of the population as a whole. Feitosa & Machado (2014) present a case in Vitória, Espírito Santo, Brazil, where a school collects around 400 mobile phones from students every day at the entrance and returns them at the exit. The reason given is that mobile phones disrupt the performance of other students and teachers. According to the school, without the use of mobile phones, the teachers' work and the students' participation has improved. Some states, such as Paraná, have State Law No. 18.118/2014-PR, of 24 June 2014, which prohibits the use of electronic devices in classrooms for non-pedagogical purposes, in primary and secondary education, being allowed only for pedagogical purposes, under the guidance and supervision of the teaching professional. The state of Santa Catarina also has State Law 14.363, which states in its first article (Santa Catarina, 2008): "The use of mobile phones in the classrooms of public and public schools in the state of Santa Catarina is prohibited".

Pompeo (2014) explains that banning the use of electronic devices is authoritarian and retrograde; the best way is to adapt to reality and use technology in favour of the classroom. The study by Beland & Murphy (2015), investigated the impact of restricting the use of mobile devices in schools in relation to student productivity, the research was carried out with 91 secondary schools, in four major cities in England, and the results indicated that there is an improvement in student performance of only 6.41% in schools that have introduced a ban. And Guenaga et. *al.* (2012) emphasises that instead of banning the use of mobile devices in education, it's better to create ways of exploiting their resources.

In order to realise the benefits of mobile learning, UNESCO (2014) recommends that policymakers take actions such as: Creating or updating policies regarding mobile learning; Training teachers on how to advance learning through mobile technologies; Providing support and training to teachers through mobile technologies; Creating and improving educational content for use on mobile devices; Expanding and improving connectivity options; Developing strategies to provide equal access to all; Promoting the safe, responsible and healthy use of mobile technologies; Using mobile technologies to improve communication and educational management; Raising awareness about mobile learning. Table 19 shows the difficulties encountered in using mobile technologies in education.

Table 19. Presents the difficulties in using mobile technologies in education.

Difficulties in using mobile technologies in education
Not sure if it's correct and suspicious of the validity and accuracy of the source of information researched
0 students being unable to get answers or solutions to their questions or doubts
Disrupting the performance of other students and teachers
Attention deficit and distraction
Tiredness when used for a long time
Lack of battery and sufficient bandwidth
Teacher unpreparedness, lack of ability to incorporate the use of mobile technologies in the classroom
Lack of planning
Lack of well-funded, coherent, large-scale and highly visible actions
Ensure that the projects take into account the realities and limitations of the infrastructures that already exist for education and ICTs, as well as the social and cultural contexts of the different countries and regions,
Cheating on exams

According to Bedi (2014), the teacher's role goes beyond simply providing new information, but also in helping students engage in the process of abstraction and differentiation of important and reliable information. Table 20 shows possible strategies to make teachers better prepared to use mobile technologies.

Table 20. Presents possible strategies to make teachers better prepared to use mobile technologies.

Strategies to make teachers better prepared to use mobile technologies
Technology should not be thrown into the classroom expecting good results, the teacher needs to focus on pedagogy, increasing the use of technology
Reducing lectures, because mass teaching doesn't work for this generation
Empowering students to collaborate, encouraging them to work together and showing them how to access experts on a particular subject available on the *Internet.*
Focusing on learning for life, not just for an exam, i.e. teaching how to learn and not what to know
Use technology to get to know each student, building appropriate learning programmes for each one
Creating educational programmes with options for customisation, transparency, integrity, collaboration, fun, speed and innovation in project-based learning experiences
0 teachers need to recognise their own digital habits and use their own digital tools in their daily lives
Reinventing yourself as a teacher

According to Tapscott (2010, p. 180), to reduce lectures, you can start by asking students questions and listening to their answers. You can also listen to the students' questions and let them discover the answer, thus creating a learning experience together with the teacher. For Batchelor, Herselman and Traxler (2010), educators are key to incorporating technology into teaching and learning. Castells (2003) says that before changing technology, rebuilding schools and retraining teachers, it is necessary to create a new pedagogy based on interactivity, personalisation and developing the autonomous capacity to learn and think.

Royle, Stager & Traxler (2014) state that teachers first need to recognise their own digital habits and use their own digital tools in their daily lives, and then use these tools for learning purposes. It was observed that the topic is being addressed more by technological areas than social and humanistic ones. It was also noted that the publications were associated with more than one area of knowledge, due to the multidisciplinary nature of the topic. Table 21 shows what still needs to be better explored about mobile technologies.

Table 21. Shows what still needs to be better explored about mobile technologies.

What remains to be better explored
Find a more precise understanding of the attributes of wireless computing that fulfil educational requirements
Providing richer pedagogical practices thanks to simple mobile technologies
Providing a standardised common platform that overcomes the tendency to fragment the market into countless different devices
Intensify and deepen research into the educational uses of mobile technologies, the requirements, methodologies and consequences of using new resources in the teaching and learning process.
Investigating new educational practices
Explore new technological resources
Research into digital literacy and literacy among teachers

In order to answer the questions posed by this research, an interdisciplinary systematic review was carried out which identified that mobile learning *(M-learning)* is an emerging and expanding field of research, due to the increasing mobility of today's society. As such, the majority of students and teachers are making use of mobile communication devices. This use involves access to teaching materials, access to environments for interaction between students and teachers, access to school activities and file sharing. Various studies have shown that mobile communication devices are an alternative for broadening students' possibilities, providing new ways of constructing and developing knowledge.

However, in order to be successful in conducting activities involving mobile technologies, it is important that teachers are properly trained and able to develop dynamic and motivating activities, making mobile communication devices useful and innovative tools in the teaching and learning process, and not simply reproducers of content. This study was also able to identify and present the benefits and difficulties of using mobile devices, possible strategies to make teachers better prepared to use mobile technologies and what remains to be explored on the subject.

Thus, it was possible to verify some of the difficulties encountered in using equipment for educational purposes, such as being distracting and disrupting students' concentration on school tasks, or the enormous benefits for education that the use of mobile devices can bring, such as allowing students to learn at any time and place or allowing the socialisation and inclusion of people with disabilities or restricted mobility, adapting to people's daily lives. Further studies involving other factors and samples and defining different inclusion and exclusion parameters are suggested to confirm the results found.

3.3. Requirements for *M-Learning* Activities

Moura (2009, p.39) defines *mobile learning* or *m-learning* as an expression used to designate a new educational "paradigm" based on the use of mobile technologies. As such, it is possible to call *m-learning* any form of learning through any small format device, with an autonomous power supply and small size, which can be used by people anywhere and at any time.

Educators are currently faced with the challenge of finding teaching tools that can be used in the classroom as interactive methodologies, making the educational environment increasingly digital (Prensky, 2012). Gómez (2015, p. 29) emphasises that "we need to reinvent school so that it can develop knowledge, skills, attitudes, values and emotions".

The Unesco report (2014) outlines the benefits of using mobile learning, including facilitating individualised learning, providing immediate feedback and assessment, ensuring productive use of classroom time, supporting learning outside the classroom, creating a bridge between formal and non-formal learning, extending education to different locations, facilitating students with disabilities, enabling learning anywhere and anytime, building different learning communities and improving communication between students and between students and teachers. According to Sharples et. *al.* (2007) mobile learning in no way replaces formal education, it simply offers support inside and outside the classroom, for the conversations and interactions of everyday life.

For Vavoula & Sharples 2002, we learn across space as we take ideas and learning resources acquired in one place and apply or develop them in another. We also learn through time, revisiting previously acquired knowledge, ideas and strategies in a completely different and broader context. When we move from one topic to another, we manage a variety of learning, rather than following a single curriculum.

CHAPTER 4

EXPLORATORY STUDIES

This chapter describes some exploratory case studies that used the concepts of the flipped classroom combined with some mobile technology. These presented significant characteristics, such as: interactivity, mobility, teamwork, learning in real contexts, greater collaboration, among others.

4.1. Exploratory research: gathering indications

This section presents an exploratory investigation into the possibilities and potential of using the Flipped Classroom Theory with the aid of Mobile Technology, through the use of *WhatsApp* for prior study and interactive activities in the classroom.

- **Materials and Methods**

The study sought to analyse the limitations and possibilities of this learning, analysing the interest of students and teachers, using technical procedures based on the action research method, since the researcher had a collaborative role in the research, promoting interventions that led to the creation of the *WhatsApp* group.

Exploratory research "tends to address new problems about which little or no previous research has been done" (Brown, 2006). Furthermore, it should be noted that "Exploratory research is initial research, which forms the basis of more conclusive research." (Singh, 2007).

A *WhatsApp* group was created which was used as: discussion forums, centres for questions, the development of collaborative texts and the sharing of links, videos, websites, images and audio that could help and stimulate learning.

This stage is characterised as action research, according to Thiollent (2011, p.20), which is "a type of empirically based social research, designed and carried out in close association with an action or the resolution of a collective problem, in which the researchers and the participants representing the situation or problem are involved in a cooperative or participatory manner". The analysis and evaluation was based on the performance of the students and teachers on *WhatsApp,* to see if there had been any changes in behaviour, acquisition of skills that would enable learning and teaching of the subjects to evolve or not.

Content analysis was used to analyse the data collected from open-ended questions. Each answer was read more than once, coded and a frequency table was created. Themes were identified and, finally, the harmonisation of codes and themes was examined. Significant statements from the participants were included as quotes to illustrate.

The data was analysed using *WhatsApp, taking into* account whether there had in fact been an improvement and increase in the level of participation in face-to-face classes, changes in attitude and commitment to studying, changes in the teacher-student relationship and an increase in school performance from one term to the next, using the term before the intervention as a parameter for comparison.

The students accessed and made the content available in a *WhatsApp* group, which was used as a learning tool. Afterwards, by applying an electronic questionnaire to the students, 44 responses were obtained from the students of the Project Management and Information Systems Projects subject of the Information Systems (IS) course at the Federal Institute of Goiás (IFG). The form used was created in *Google Docs.* We also used notes from the teacher's diary, observation and face-to-face interviews carried out at the end of the intervention programme.

• Results and discussions

This section presents the results and experiences gained from the proposed experiment. The project lasted eight weeks, which corresponded to various syllabuses of the planned curriculum. The students took an active part in exchanging messages in the group created. The students interacted quite frequently and naturally, and class engagement increased as the proposed activities developed.

The students' answers revealed that the use of mobile devices made it possible for learning to take place anywhere and at any time. The students replied that they studied at home, at work, in their free time, during lunch breaks or between classes, making their study schedules more flexible. All the students said they had a smartphone and used it to carry out their research and studies.

The majority of students (93.2 per cent) reported that they had not encountered any difficulties, while the remainder (6.8 per cent) said that they had struggled to understand the content, requiring prior, face-to-face explanations, as shown in Table 22.

Table 22. Difficulty encountered in understanding content

Found it difficult to understand the content	
Yes	6,8%
No	93,2%

They said they had no difficulties with the size of the mobile device screen, the resolution of the display, reading information or watching videos on mobile devices. Table 23 shows the positive aspects of students' learning using mobile devices.

Table 23. Positive aspects of learning

What are the positive aspects for your learning?
Sharing experiences
Knowledge and content
Flexible working hours
Practicality of having content in various formats such as videos, slides, audios, etc.
Integration between classmates

When asked how well they had learnt the activity, on a scale of 1 to 5, with 5 being the maximum, 11 students (25%) answered 5, 15 students (34%) answered 4 and 18 students (41%) answered 3. None of the students answered 2 or 1, as shown in Table 24.

Table 24. Level of learning from the activity

How well did you learn from the activity?				
1	2	3	4	5
0%	0%	41%	34%	25%

During the interview, one student made a suggestion to improve the discussion and, consequently, learning, by setting a time for everyone to be online to facilitate instant debate between all the students.

Table 25. Level of student satisfaction

Level of satisfaction	**Very satisfied**	**Satisfied**	**Dissatisfied**	**Very dissatisfied**
Proposed activity	40,9%	54,5%	4,5%	0%
Way of communicating	34,1%	54,9%	11,4%	0%
Group created	45,5%	50%	4,5%	0%
Screen size	38,6%	59,1%	2,3%	0%
Methodology used	31,8%	61,4%	6,8%	0%
Application used	38,6%	61,4%	0%	0%

As shown in Table 25, the students were asked how satisfied they were with the proposed activity. They answered: 54.5% satisfied, 40.9% very satisfied, 4.5% dissatisfied. Asked how satisfied they were with the way they communicated with their classmates and teacher? They answered: 54.9% satisfied, 34.1% very satisfied, 11.4% dissatisfied. What is your level of satisfaction with the group you created? They answered: 50% satisfied, 45.5% very satisfied, 4.5% dissatisfied. Asked what their level of satisfaction was with the suitability of the content for the size of the screen? 59.1% satisfied, 38.6% very satisfied, 2.3% dissatisfied. Asked how satisfied they were with the methodology used in the activity? 61.4% satisfied, 31.8% very satisfied, 6.8% dissatisfied. Also asked how satisfied they were with the application used in the activity? They answered: 61.4% satisfied, 38.6% very satisfied, 0% dissatisfied. No students said they were very dissatisfied.

Asked what their opinion was on the use of smartphones in the subject? They replied:

- "I think it's a good thing, since all the students have a mobile phone, and the mobile phone is a computer and will help us with our research and so on."
- "I find it very interesting, a new way of teaching."
- "I think it's a good idea because of the greater interaction between student and teacher."
- "Incredible, I find the teaching engaging, which helps to assimilate the content."
- "It's a way of connecting all the students in an interactive way, I think it's an interesting activity."
- "Very interesting as it makes the lesson more dynamic.";
- "A different and didactic way of teaching.";
- "Fun and didactic, but it does hold my attention a little when it comes to content due to notifications from social networks and so on."

Asked about the potential of using mobile technology in the learning process? They reported:

- "Proximity to the student and practicality";

- "It makes the learning process easier."
- "Good, technology helps our performance a lot.";
- "The world evolves, so do we! I use my mobile phone more than my notebook."
- "It improves interaction with the students and the teacher."
- "It mixes something we like to do (use a mobile phone) with what we have to do (study)."
- "It's invaluable, because as well as helping with learning, it's fun and really cool."
- "The use of mobile technology allows us to learn in a more interactive and attention-grabbing way, so it becomes a good tool to use."
- "The use of technology can increase student interest if used correctly."
- "It increases communication, learning and didactic diversity.".

Asked what are the limitations of using mobile technology in the learning process? They reported:

- "Lack of internet";
- "It can disperse a lot."
- "Technology is not enough for learning, but rather a complement."
- "That wasn't the case, but some students might not have mobile phones."

Among the limitations pointed out were financial and technical problems, because it could exclude some students who don't have smartphones, internet plans on their mobile phones or internet at home.

Asked what are the positive points of using mobile technologies in the learning process? They answered:

- "Greater interaction between people and practicality";
- "An easier way to do research, communication, etc."
- "Get the student interested";
- "Easier learning, easier communication";
- "And a faster way to communicate and study.";
- "The dynamics and interest of the student are greater, the relationship between student and teacher is more interactive."

- "0 mobile phone is much faster than a computer.";
- "More familiarisation, accessible technology, fun and didactic".

The students' responses showed that 89.3 per cent found the pedagogical activities productive and only 10.7 per cent felt they were not, mainly due to the fact that they didn't feel totally comfortable expressing their doubts through group messages, as shown in Table 26.

Table 26. Productivity when using WhatsApp

WhatsApp p productive in teaching activities	
Yes	89.3%
No	10,7%

It was possible to see an improvement in the quality of the learning process in the subjects, measured by the results achieved through an increase in the average grades of the classes. In one, in the first term, the class average was 4.6. In the second term, with the new pedagogical proposal, the average rose to 7.5. In another, the class average was 5.8. In the second term with the new pedagogical proposal, the average increased to 8.2. As Table 27 illustrates.

Table 27. Average grades in subjects

Increase in grade point averages		
*	First academic term	Second term
Subject 1	4,6	7,5
Discipline 2	5,8	8,2

We can cite a few more positive points reported by the students with the use of *WhatsApp: it* improved the relationship between the teacher and the students; it made studying easier; it made classes less boring; it provided more time to study; it helped to clear up doubts; it allowed students to study more and help their classmates; it made it easier to study subjects that they didn't understand much just with the teacher's explanations in class; it helped to clear up doubts without having to wait until the next class; increased communication with the teacher; made it possible to study anywhere and always have the teacher close by; provided collective, collaborative learning and constant exchanges of knowledge; made it possible to advance content and subjects; allowed the use of videos, audios and images to better illustrate the content to be studied; allowed the creation of groups for socialising. It was even suggested that each teacher should create a group for their subject. An important fact reported is the closer ties between teachers and students, because according to the students themselves, the lessons became "less boring" and the students felt that the teacher was more present and active inside and outside the classroom.

There was also an improvement in student behaviour and in the search for more advanced knowledge in relation to the subject content. One point to highlight was the huge amount of study material in various formats that was made available in the group. One of the main resources used by students were audios with explanations or contributions on the content. Many links were also posted, making it easy to find material on the topics covered. Another fact was the occurrence of repeated questions, since some students didn't follow the group's history and ended up asking the same thing as other colleagues. This intervention confirmed other studies that have also pointed out the pedagogical potential of mobile technologies, such as: Rambe & Bere (2013) - interactivity, knowledge sharing, motivation, collaboration, synchronicity and asynchronicity; Park, Cho & Lee (2014) - feeling of presence, sharing emotion; and Padrón (2014) - low investment.

• .1.3 Considerations

The research showed that at the end of the activities the students felt very motivated and active in the learning process, as they made choices about which materials, how and when to access the available content. Observing the activities carried out by the students, it was possible to see that there was a great effort on the part of the students to achieve the proposed objectives, proving the motivating aspect of the activity. Another interesting point was that some students chose content that wasn't made available by the teacher, demonstrating great flexibility and making it possible to grow in autonomy within the context. It was also possible to observe, by analysing the answers, the maturity of the students in relation to responsibility, many of whom questioned whether the tool could be a means of distraction and thus harm their learning.

During the course of the activity, various practical difficulties were exposed in relation to the subjects, which were clarified by the teacher, consolidating and honing these concepts in the students in an iterative way outside and inside the classroom. Another important point was that some students were able to glimpse other, non-trivial areas of research, demonstrating an improvement in the search for other knowledge.

The students felt comfortable using the app for teaching purposes under the supervision and planned guidance of the teacher.

The activity allowed the students' development to mature within the context of learning, interacting with multiple aspects within a learning axis. Another relevant finding was that self-taught practice can be developed in students through this type of activity, broadening students' investigative nature, as well as their interest in seeking help. The main conclusion that can be drawn is that the use of inverted classroom approaches combined with the use of mobile technologies strongly solidifies learning and aims to ensure greater use of concepts. The results have led teachers of other subjects to adopt this pedagogical methodology in different classes, with excellent results.

One of the limitations and difficulties that can jeopardise the goal of making mobile applications an extensive and complementary part of the classroom is financial and technical problems that can exclude some students who don't have *smartphones, internet* plans on their mobile phones or even *internet* in their homes, making it difficult for them to use and especially access the tools and digital resources available.

Another factor was closer ties in the teacher-student relationship, because, according to the students, respect and consideration between students and teacher increased as they got to know each other better. The distance was reduced, making lessons "less boring", and the students felt that the teacher was much more present and active inside and outside the classroom.

Among the most significant benefits found are: low cost, accessibility, interactivity and collaborative learning. Some of the negative aspects include the possibility of a lack of access to technology and the difficulty teachers have in adapting their teaching practices to the new technologies and means of communication. On the other hand, it was clear that mobile technologies, when used in education, require a mediator, and that the great advantage of applications such as *WhatsApp* is the speed with which questions can be viewed, which can be answered by both the teacher and the classmates themselves, and especially the confirmation that messages have been viewed, confirming the study by Alencar *et. al.* (2015).

The students appreciated the possibility of sharing materials, instant messaging, the opportunity to upload files, discussions and receiving instant notifications. All the students approved of the communication and discussion that the technology provided between the students and with the teacher. All the students were in favour of using mobile technologies in education. The vast majority of students, as well as the class teachers, stated that the methodology facilitated the promotion of teaching and learning. The results of the intervention confirmed the references found in the literature.

4.2. Pilot Case Study 1: Using the *Facebook* Social Network as a Teaching Support Tool

4.2.1. Contextualisation

Internet use is growing rapidly these days, especially among young people. People all over the world have the opportunity to share experiences and information. According to the United Nations Children's Fund (UNICEF), more than 10 million adolescents use the *Internet* every day around the world, with one of their main activities being access to social networks, entertainment and searching for information (Unicef, 2013). According to the Brazilian *Internet* Steering Committee (2014), the 2013 ICT Households survey carried out in Brazil showed that 75 per cent of teenagers aged 10 to 15 and 77 per cent of young people aged 16 to 24 use the *Internet. This* shows that educational practices can be favoured through the use of DICTs, especially social networks such as *Facebook* (Seabra, 2010).

Facebook defines itself on its official website as a product/service whose mission is to "create a more open and transparent world, which we believe will create more understanding and connection." *(Facebook,* 2018). Currently considered the largest social networking site in history and the most powerful means of communication of our time, according to issue 348 of June 2015 of Super Interessante magazine, 936 million people log on to *Facebook* every day, 59 million of them in Brazil alone. According to the same magazine, "Half of all people with *Internet* access in the world log on to *Facebook* at least once a month." (Santi & Garattoni, 2015).

There are studies that point out that *Facebook* can be used as a VLE, as it brings together various types of media in a single environment, enabling and providing opportunities for collaborative learning, interactivity and various pedagogical practices that lead to learning how to learn (Ferreira; Correa & Torres, 2012). However, teachers in Brazilian schools are still resistant to the inclusion of technologies in the classroom, as they are still stuck in traditional pedagogical practices, and are unable to include new technologies in the school (Lima, Andrade & Damasceno, 2010).

The aim of this case study is to report on the experience of using *Facebook* as an aid to teaching and learning in a face-to-face higher education course. It thus shows the possibilities of an educational intervention mediated by the use of this social media in learning. The target audience for the experiment was students on the Project Management subject of the Higher IS course at IFG Câmpus Luziânia.

This study is justified because education professionals are increasingly discussing the use of social networks in the educational context. Thus, this research seeks to contribute to the possibilities and potential of *Facebook* in the classroom and beyond. According to Ferreira, Correa & Torres (2012), advances in the applicability of innovative methodologies combined with Web 2.0 are extremely necessary so that teachers can improve their pedagogical practices in the face of students connected to TDIC.

4.2.2. Literature review

Facebook is currently considered a worldwide phenomenon, given its high visibility. The social network is increasingly representing a new way of establishing relationships, carrying out tasks such as publicising products, news, facts, sharing videos, texts, ideas, photos, images and fun through its applications (Ferreira; Correa & Torres, 2012).

The social network *Facebook* was created in 2004 with the aim of interacting with people and sharing information, images and videos, and is currently one of the most widely used. Social networks are used for entertainment, but they can also contribute to better access to information, education, social and political intervention. These social networks are part of the daily life of Generation Z, who are people born after 1993 and are also known as digital natives, as they have been using digital media since birth. Nowadays it's a difficult task to live disconnected from the *internet.* Thus, the use of TDIC in education has been growing in recent years and can be a pedagogical facilitator. These media can contribute to interactivity in the classroom (Alencar; Moura & Bitencourt, 2013). Mattar (2012) "supports the idea that social networks have the potential to generate interaction, as they unite people with common interests".

Social network groups can function like a traditional study group, where students and teachers share useful information that will help with classroom activities. It's a way for students and teachers to work on collaborative projects, and it's possible to create open and closed groups, helping to preserve students' privacy. Chat can help exchange information directly between teachers and students, and can facilitate interaction with teachers beyond the school walls. The mural, according to Mattar (2012), can be a space for communication and discussion, where teachers can encourage student participation. Events can be used to remind students of certain deadlines for handing in work, meetings and lectures (Alencar; Moura & Bitencourt, 2013). Table 28 illustrates how *Facebook* tools can be used to support teaching.

Table 28. Facebook tools in teaching. Source: Juliani et. al. (2012)

Tools	**How to use it?**
Chat	Ask questions in real time. Teacher and Teacher, Student and Teacher, Secretary and Student, Community together with students, teachers and secretary.
Photos and Videos	Publicise the work and activities carried out. For example, a video of a lecture held on campus, or photos of a field study. It's important to look for the best quality image to publish.
Sharing	Disseminate relevant information and knowledge to *Facebook* users who do not participate directly in the groups created (curricular units/disciplines)
Events	Publicise and receive confirmation of participation in meetings, trips, lectures, etc.
Comments/Message	Remind students about exams, assignments and resolve individual doubts. Create an atmosphere of interaction/debate on certain topics.
Polls	Gathering the opinion of students or other stakeholders on a given subject.
Contents	Creation of new pages within a group. Different subjects can be posted and stored indefinitely. Examples: Exam notes, lesson summaries, teaching plans.
Tagging images, videos and comments	Whenever possible, mark all those involved in the content to make it clear and encourage participation.
Debates	When the teacher publishes material, it is also possible to publish a space for debate on the subject, advising students to leave just one comment, and then debate the subject with their classmates and teachers in order to better grasp the content.

Facebook has been occupying an important space in education as a platform for communication. Mattar (2013, p. 115) points out that relationships between teachers and students via *Facebook* generate "a more open communication channel, resulting in richer learning environments and greater student involvement in schooling processes". This social network allows teachers to use different methodologies to encourage and motivate students in the learning process (Ferreira; Correa & Torres, 2012).

When *Facebook* is used as a resource or VLE in teaching, it allows the teacher to modify the way of learning in a more interactive and participatory way, facilitating pedagogical mediation and interaction.

This network makes it possible to incorporate, personalise, resize, streamline and add meaning to learning, thus making it more attractive, allowing students to move away from the role of simple passive recipients and become responsible for their own learning (Ferreira; Correa & Torres, 2012).

Llorens & Capdeferro (2011) pointed out the main pedagogical potential of *Facebook* for collaborative learning: favouring a culture of virtual community and social learning, allowing innovative approaches to learning and enabling the presentation of content. The tool proved to be efficient in promoting collaborative learning, fostering critical thinking, providing debate on the content presented and the diversity of knowledge, favouring collaborative learning and the exchange of knowledge experiences.

However, social networks were not created for educational purposes, although they can be used as VLEs. It is therefore up to teachers to select and problematise information in order to teach and learn. Social networks offer teachers great pedagogical potential and countless educational possibilities. It is up to the teacher to know how to use *Facebook* as a VLE, favouring collective, interactive and contextualised learning (Ferreira; Correa & Torres, 2012).

The big disadvantage of using *Facebook* instead of other virtual learning tools is evaluating the content posted by students, as this is not a tool created specifically for educational purposes. However, a statistical and qualitative analysis of students' participation (posts) can be carried out manually or using social network monitoring tools. Using this type of automated tool, it is possible to evaluate each student's contributions and assign a concept to their participation (Juliani et. *al.,* 2012).

According to Seabra (2010), for these technologies to be meaningful, it is not enough for students simply to access the information; they need to use it, relate it, synthesise it, analyse it and evaluate it. The teacher has to be a mediator, challenging ideas and conclusions. Seabra (2010, p.20) also states that "The use of social networks in the educational process must be done in a well thought out way, as it risks being just a distraction, generating more noise than helping the teaching and learning process.". Thus, teachers can underutilise or even distort the potential of social networks, instead of using them as a teaching and learning tool.

4.2.3. Materials and Methods

The research presented here takes a quantitative approach and is characterised by information gathering, which is the direct questioning of people whose behaviour we want to know, requesting information from a group of people about the problem. A questionnaire was used for this purpose (Gil, 2002). The case study was the result of an assignment proposed by the Project Management teacher on the IS course at IFG Câmpus Luziânia.

The experiment involved the participation of twenty-six students and a mediator, the subject teacher, who proposed topics related to the subject for discussion *online* for a period of one week, in a subject group on *Facebook,* functioning as a discussion forum, similar to Moodle, but more familiar to the students, providing more interactivity and portability, due to the possibility of accessing it via *smartphones.* After the suggested debates, an *online* questionnaire was sent to the participants, in order to profile them and gather information to compose the research results. The answers were tabulated using Excel software and then analysed.

4.2.4. Results and discussions

Of the participants who answered the questionnaire, six were female and 20 were male. 69.2 per cent were aged between 18 and 25, 23.1 per cent between 25 and 30 and 7.7 per cent over 30. The identities of the participants have been preserved. The answers provided in the questionnaire revealed some relevant information, as shown in Table 29.

Table 29. Students' questions and answers.

Questions asked	Yes	No
Do you use *Facebook* every day?	100%	0%
Are teachers part of your *Facebook* contacts?	76,9%	23,1%
Have you ever questioned a teacher on *Facebook?*	46,1%	53,9%
Can *Facebook* be used to support teaching?	92,3%	7,7%
Were you satisfied with the use of *Facebook* in your course?	92,3%	7,7%
Did the use of *Facebook* in the course allow for the exchange of information between students and between students and teacher?	84,6%	15,4%
Was the use of *Facebook* in the course important?	92,3%	7,7%
Did the use of *Facebook* in the course help to clear up doubts?	96,1%	3,9%
Was the use of *Facebook* in the course effective?	92,3%	7,7%

Analysing the data in Table 29, it can be seen that the use of *Facebook* is part of the students' daily lives, and all the students said they were part of the network, which is precisely why it is important to think about new proposals that include this tool in educational activities. According to Pereira *et al.* (2012), when used with specific and well-defined objectives, *Facebook* can promote interaction and help in the teaching-learning process. Most of the students have their teachers in their *Facebook* contacts, but less than half of them have ever asked any of their teachers about a subject.

Most of the participants said that *Facebook* can be used to support teaching, and they were satisfied with the use of *Facebook* in the subject, especially since it is an important tool for communication and sharing content. Its main advantage is that it is easy to exchange messages synchronously when they are *online* or asynchronously when they are not. This type of communication is important because the student can ask the teacher questions at any time, without the need for personal contact, thus collaborating with the teaching and learning process, since the exchange of information is almost in real time. *Facebook* can be a collaborator in teaching, but more initiatives are needed to use it.

The majority also said that the use of *Facebook* in the course enabled the exchange of information between students and between students and the teacher, in various formats, such as videos, audios, slides, among others, and that the use of *Facebook* in the course was important, as it helped most students to answer their questions. Showing that the tool can be very effective as a support in pedagogical practices, adding value to the process of teaching and learning content.

Moran (2013) says that technologies affect education, changing the place and time of study, but in the meantime there is still a lot of resistance, mainly due to the lack of training for professionals to use TDIC and the fact that these resources can cause students to become distracted.

It was also observed that the use of mobile technologies facilitates interaction and breaks down the barrier imposed by temporal spaces. As Saboia, Vargas & Viva (2013) pointed out. Currently, the use of

Traditional teaching methods no longer make sense, given the ease of access to information and technology, such as the use of mobile devices and *Facebook* in the school environment.

The students commented that the practicality and ease with which they have access to *Facebook, and the fact* that it is widely known and used among them, also facilitated learning in a relaxed, enjoyable and dynamic way, as well as contributing to increased autonomy, seeking more knowledge than requested by the teacher, allowing the students to become responsible for their own learning. They felt motivated to take part in the discussions because they could ask questions at any time. Some reported losing their shyness using the platform and thus participated more actively in the discussions, although one student said he didn't feel comfortable expressing his opinions via the *Facebook* group.

Student participation proved to be effective, providing links to texts, videos and slides. Most of the students said that using the social network facilitated discussions, making it easier to interact *online* than in person, and that they felt more comfortable communicating than in the classroom. Some students reported that they liked using *Facebook* because even when they missed classes they could access the content they had missed, because it was available *online, and they* could review it as many times as necessary to learn. It also favoured the bond between the students and between the students and the teacher.

Thus, according to Aragão *et al.* (2018), educational strategies should be dynamic, participatory, interactive, playful, meaningful and joyful. The use of social media, such as *Facebook,* favours the achievement of these objectives, as they are part of students' daily lives, as long as it is planned.

4.2.5. Considerations

This report has shown that *Facebook* can be an excellent tool used in the educational context. It was observed that the VLE can be replaced by *Facebook,* or simply serve as an alternative to this environment, functioning as a forum, sharing content related to the subject and solving doubts with the help of colleagues and the teacher. It was found that mediation by the teacher is fundamental in this type of activity, proposing themes and encouraging student participation. Another strong point in the teaching and learning process is the possibility of communication and interaction between students and with the teacher in a wide variety of places or at any time the student feels it necessary, especially via a mobile phone.Thus, it can be concluded that *Facebook, in* the opinion of the participants in this study, can be a very effective tool to support teaching practices, adding value to the process of teaching and learning content. Analysing the results showed that the students support the idea of using *Facebook* as an educational platform.

4.3. Pilot Case Study 2: Using *WhatsApp* as a Learning Support Tool

4.3.1. Contextualisation

Instant communication technology, mediated by mobile phones and *smartphones, has* completely changed the way people communicate and relate to each other, and educators cannot fail to consider this equipment for use in the educational process, especially as an integrator between those involved (Pereira; Pereira & Alves, 2015). The use of TDIC favours the dynamisation of teaching and the production of new scientific and cultural knowledge. As such, various initiatives are being developed to promote practices with these tools in order to help improve the quality of teaching and learning (Borges, 2015).

Gomez (2010, p. 88-99) states that: "Social networking programmes, whether personal, thematic or professional, were not actually created for educational activities."*The* same author also states that the pedagogical use of *online* social networks and applications for mobile devices requires a certain amount of care, saying that: "The web is yet another space in the contemporary school that needs guidance and care if it is to become a pedagogical device."

The use of digital tools in the school context simply does not guarantee quality in education and requires a transformation of teaching practices as a whole. Moreira & Kramer (2007, p. 1038) state that: "In short, it is as if technical objects could magically guarantee quality in education. In many cases, the content traditionally taught in the classroom is transposed into new media".

In this context, the use and promotion of *WhatsApp* as an extension of the classroom can provide an environment for learning and collaboration. It can be an alternative for the pedagogical use of mobile devices without having to prohibit it, and can make lessons more attractive and enjoyable, increasing the possibilities of performance and learning, breaking the spatial and temporal boundaries of the classroom (Lopes & Vaz, 2016).

The use of *WhatsApp* in education is in line with Moran's (2013) assertion, as it allows students to be connected to the virtual world and communicate via mobile devices with other students and teachers. At this point, the teacher takes on the role of advisor, facilitator and mediator of knowledge and the student as a builder of their own knowledge, through research and questioning regarding doubts that arise, making learning much more meaningful.

In recent years, it has become easier to access mobile technologies, which means that they need to be gradually introduced into school activities. They are gradually adapted by teachers, students and the school environment (Alencar *et al.,* 2015). Schools, especially Brazilian teachers, are resistant to inserting new technologies into the classroom because they are involved in a world of traditional pedagogical practices (Lima, Andrade & Damasceno, 2010). The aim of this case study is therefore to analyse the possibilities and potential of using *WhatsApp* in secondary education, with a view to helping those interested in using it in teaching.

4.3.2. Bibliographical review

This section contextualises mobile technologies, *WhatsApp and* ICT in teaching and learning processes. *WhatsApp* is a multi-platform application that uses the *internet* to send and receive instant messages free of charge and without limits. The application makes it possible to send different media, such as images, audio and videos. There are other important features, such as the possibility of creating groups with up to 100 members, transmitting dialogues and making calls. One of the main advantages of this app is that it synchronises with your contact list, so you don't need to remember your username and password, just add or have the numbers saved in your mobile phone contacts (Alencar *et. al.,* 2015).

WhatsApp is a tool used by a large proportion of students, if not all of them, and can facilitate the communication process between students and teachers, as well as between the students themselves, creating a favourable scenario for debates on various subjects (Paiva et *al.,* 2016). Honorato & Reis (2014, p. 3) state that "for students, the advantages of the *WhatsApp* app are that they can pass on information about subjects, ask questions about content, tasks or assignments".

According to Niza (2016), the use of *WhatsApp* in schools can facilitate sharing between team members, as the groups created on the app allow teachers to exchange experiences about their teaching practices and recommend teaching materials and activities, as well as bringing parents closer to their children's school routine by sending messages, It also allows extra content and activities to be made available to students, functioning as an AVA, making it possible to make content available in audio and video format and to set up discussion forums and question centres.Oliveira et al, (2014, p. 3484) states that "*MLearning* is understood to be the combination of the concepts of mobility and learning". In this sense, Moran (2013) says that schools have an obligation to make use of technology, and take advantage of it. Nowadays, it is unacceptable for teachers not to use technological resources in their lessons. Coscarelli & Ribeiro (2011, p. 91) say that we need to recognise the importance of technology, making it an ally in the production of knowledge. Moran (2013, p. 27) also states that the school needs to play an innovative role, inserting the *internet* as an important methodological component. For Moran (2013, p. 31), the use of ICT in the classroom can bring significant improvements to learning, transforming the school into a rich learning space, motivating students to learn actively, to research all the time, to be proactive, to know how to take initiatives and interact". According to Lévy (2010, p. 173), the teacher's role can no longer be to disseminate knowledge, which has already been made available effectively by other means, but to encourage learning and thinking. The teacher has now become a learning manager, encouraging the exchange of knowledge and mediating the learning process. The teacher's role, according to Moran (2015, p. 24), is that of curator and advisor, choosing what is relevant from the countless pieces of information available and guiding students with the materials and activities available, whether in groups or individually.

4.3.3. Materials and Methods

This section presents the methodology used, which was a case study, with the instruments: direct observation of interactions and students' perceptions. The research was applied to thirty-two students on the web authoring course of the technical course in *internet* computing at IFG Câmpus Luziânia.

The students accessed and made the content available in a *WhatsApp* group, which was used over four weeks as a learning tool. After the debates, an electronic questionnaire was created in *Google Docs and* sent to the students to collect information and compose the results, obtaining thirty-two responses. The group was used as a discussion forum, a place to ask questions, to develop collaborative texts and to share links, videos, websites, images and audio that could help and stimulate learning.

The research was carried out with an exploratory and descriptive approach, using technical procedures based on the action research method, since the researcher had a collaborative role in the research, promoting interventions that consisted of creating the *WhatsApp* group, defining the themes and monitoring the posts. According to Thiollent (2011, p.20), action research is "a type of empirically based social research, designed and carried out in close association with an action or the resolution of a collective problem, in which the researchers and the participants representing the situation or problem are involved in a co-operative or participatory way". The analysis and evaluation was based on the participation of the students and teacher on *WhatsApp*.

4.3.4. Results and Discussions

The answers collected from the electronic questionnaire showed that all the students answered that they have a smartphone-type mobile phone and use *WhatsApp*. Like Araújo & Bottentuit Junior (2015) in their research, practically all students now have a smartphone with *internet* access. Therefore, it is important to think about proposals that include smartphones in their activities for educational purposes, given the use of these devices today. The students taking part in the research were asked if *WhatsApp* could be used as a tool to support teaching and learning, and all the students said yes.

How satisfied are you with the proposed activity?

22 replies

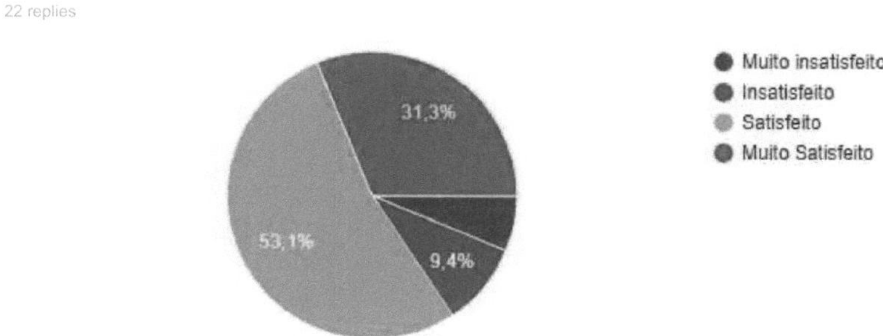

Figure 2: Level of satisfaction with the activity

Figure 2 shows that 53.1% said they were satisfied with the activity. Another 31.3% said they were very satisfied with the activity, 9.4% said they were dissatisfied and the rest said they were very dissatisfied with the activity.

How satisfied are you with the way you communicate with your classmates and teacher?

32 answers

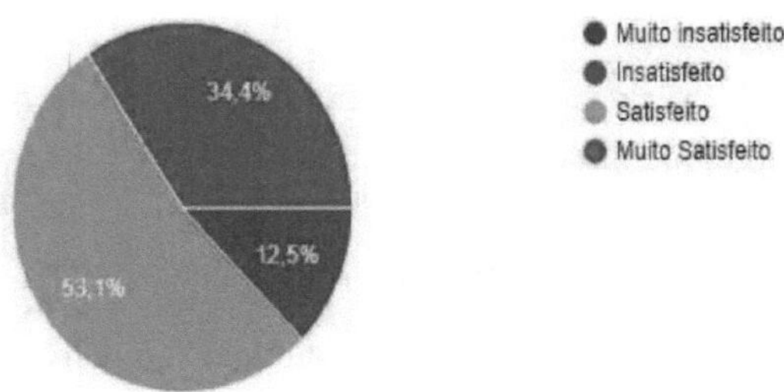

Figure 3: Level of satisfaction with the way of communicating

Figure 3 shows that 53.1 per cent said they were satisfied with the communication. Another 34.4 per cent said they were very satisfied and only 12.5 per cent said they were dissatisfied.

How satisfied are you with the methodology used in the activity?

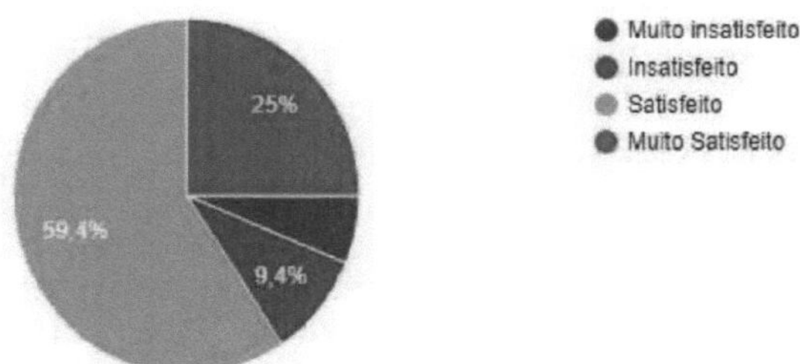

Figure 4: Level of satisfaction with the methodology used

Figure 4 shows that 59.4% said they were satisfied with the methodology used. Another 25 per cent said they were very satisfied and the rest said they were dissatisfied or very dissatisfied.

How satisfied are you with the application used in the activity?

32 resaosta'5

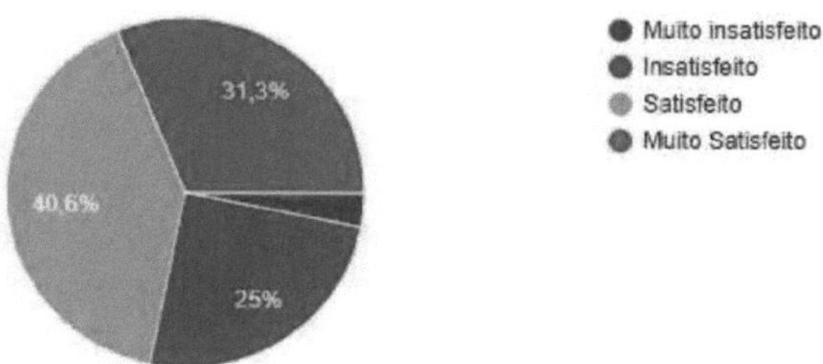

Figure 5 - Level of satisfaction with the application used

Figure 5 shows that 40.6 per cent said they were satisfied with the application. Another 31.3% said they were very satisfied, 25% said they were dissatisfied and the rest said they were very dissatisfied. Analysing the graphs, it can be seen that the majority of students said they were satisfied or very satisfied with using the app both inside and outside the classroom.

When asked how the use of the app during the activity contributed to their learning? Table 30 illustrates how some of them responded.

Table 30. Contribution to student learning.

Question	Answers
How did using the app during the activity contribute to your learning?	Using mobile phones, it was possible to interact with the class; It made it easier to access the content I was studying; I think it's cool and creative; I liked it, it brought interaction with the class and debate on the content; I don't know much, but I thought it was cool, I got more into the subject; Ease of use, because we're already familiar with it; I was able to understand the content much more easily; I was able to discuss the topic and add knowledge; It's more productive; it makes a dynamic contribution; It helped them to interact with their classmates and the teacher and to increase their interest; It increases interest in the subject; He made a good contribution.

The students reported the following as possible limitations: not having *internet* and allowing students to get distracted and use their mobile phones for other purposes. The students asked for the app to continue to be used for communication and sharing content in the subject and said they would ask other teachers to start using *WhatsApp* in their classes too.

4.3.5. Considerations

This case study analyses the possibilities and potential of using *WhatsApp* in secondary education. In this sense, it was observed that at the end of the activities the students felt very motivated and active within the learning process using *WhatsApp,* as it allowed them to share and access content as and when they wanted.

It was also noticeable that in the activities carried out by the students, there was a great effort on their part to achieve the proposed objectives, proving the motivating aspect of the activity. Another important point was that some students chose content that was not provided by the teacher, demonstrating autonomy. It was also possible to observe, by analysing the answers, the maturity of the students, since many questioned whether the tool could be a means of distraction and thus harm their learning, showing their responsibility.

During the activity, various practical difficulties were presented in relation to the subjects, which were clarified by the teacher, consolidating and honing these concepts in the students in an iterative way outside and inside the classroom. Another important point was that some students were able to glimpse other, non-trivial areas of research, demonstrating an improvement in the search for other knowledge. The students felt comfortable using the app for pedagogical purposes, but it was clear that the teacher's mediation, supervision and guidance is extremely important in order to guarantee performance and not deviate from the proposed objectives.

4.4 Pilot Case Study 3: Using the Kahoot Tool in Teaching

4.4.1. Contextualisation

Students who belong to the "digital generation", who were born in the midst of technology, have easy access to *smartphones and* other devices connected to the *internet,* for whom information is always available, and they need the teacher less and less to transmit information. Teachers who continue to want to impart all-encompassing knowledge are increasingly finding disinterested, inattentive and unmotivated students who don't see the point of going to class in search of knowledge (Tardif, 2011).

Gamification is the use of game elements outside the context of games, such as in education (Coil *et. al.,* 2017), allowing to add value to classes, providing challenge, pleasure and entertainment in the transmission of knowledge. In this context, Amico et. al. (2017) considers it important to evaluate tools such as Kahoot and other gamification platforms, as they have the potential to become consolidated trends in the field of education.

In contemporary times, various approaches and possibilities have emerged for expanding pedagogical actions in the classroom, as well as their potential (Figueiredo; Paz & Junqueira, 2015). Gazotti- Vallim (2017) states that the use of digital information and communication technologies, as well as active methodologies and gamification at school have been the subject of much discussion. These strategies have proven their effectiveness in attracting students, thus enabling an improvement in their academic, social and cognitive performance. Therefore, this case study describes an experience report of an activity developed on the Kahoot digital platform and an analysis of its potential for learning.

4.4.2. Literature review

This section contextualises Kahoot and gamification in education. Kahoot is a free game-based learning platform whose institutional mission is to "unlock the deepest potential of every student of every age and in every context" through fun, magical, inclusive and engaging learning. This platform makes it possible to create questionnaires, discussions or surveys that can be answered by users who are connected to the *internet* via *smartphones* or computers. To use it, you need to register on the virtual learning platform (https://kahoot.com/). Kahoot's aim is to engage students through quizzes, discussions and pre-prepared game-like surveys, with scoring, interaction and ranking (Dellos, 2015).

Students don't need an account to use Kahoot. To log in, they enter a pin and their surname. When they start, the questions along with the answers are shown on the large screen, and the students press the same colour and symbol with the answer they believe to be correct on the mobile screen. A timer is displayed, which slows down to zero, as well as the number of students answering the questions. At the same time, students receive individual feedback on how they answered on their devices. The students' answers provide the teacher with feedback on the students' understanding of the question, and create an opportunity for discussion about the question and the answers. The top five score, with points and nicknames, is shown between questions. Each student can also follow their own score and ranking on their own mobile device. To get a high score, students need to answer the questions correctly and quickly. Music and sound effects are used in Kahoot to create suspense and the atmosphere of a game show (Wang, 2015).

Kahoot's main idea is to be a platform where the teacher and students can interact in the classroom by simulating a competitive knowledge game. The motivation is to engage students by transforming the classroom into a game show, where the teacher would be the presenter and all students can compete by earning points through correct answers to various questions related to the subject being taught in class (Wang, 2015). Kahoot uses three gamification techniques: score, time limit and points scoreboard, but leaves out several other techniques that could be used, such as: medals, progression, long-term objectives, varied mechanics, among others (Faria; Costa & Parreira Júnior, 2016).

Games, whether on consoles, *smartphones* or computers, are widely accepted in the context of the popularisation of technology. Thus, the use of games in teaching has great advantages, such as interactivity, overcoming challenges, goals and objectives together with a high level of involvement and motivation (Pereira & Pimentel, 2014). Deterding *et. al.* (2011, p.11) defines gamification as the use of game design elements in contexts other than games. The authors also say that although the vast majority of gamification application examples are digital, the term should not be limited to digital technologies alone.

The use of gamification in educational practices, according to Kapp (2012), is defined as "the use of mechanics (rules, feedback, levels, rewards and scoring), aesthetics (game design with an attractive interface) and game thinking (building pleasant experiences, feelings of fulfilment, competition, conflict, cooperation) to engage people, motivate action, promote learning and solve problems". The principle of this strategy is to use an environment in which students can carry out their activities in a motivating way, applying elements such as: competition, exploration, co-operation and a differentiated narrative. The strategy enhances learning through a set of game elements, in order to generate involvement and dedication similar to that which games provide their players.

Gamification seeks to engage people on an emotional level and motivate them to achieve set goals. Gamified systems hierarchise the learning process, creating progressive levels of difficulty, similar to games. Thus, at each level, the student develops new skills in order to reach the next stages. Thus, learning takes place progressively, where skills and knowledge are built on what has been learnt previously (Burke, 2015). For Kapp (2012) gamification, when well designed, can help students acquire skills, knowledge and competences in a short period of time, with a high retention rate of the content.

Publications on the subject of gamification in education point to various reasons for its adoption, when well planned, the main ones being: to allow constant feedback in shorter cycles, favouring faster learning; to improve the retention and creation of knowledge; to increase the perception of individual and collective achievement; to increase engagement; to reinforce learning and development (Dicheva *et. al.,* 2015; Domínguez et. al., 2013; Morrison & Disalvo, 2014).

Lee & Hammer (2011) present the advantages and disadvantages of using gamification in education. It can motivate students to engage with the classroom, give teachers better tools to guide and reward students, encourage students to maintain lifelong learning, and can be a fun experience. Possible disadvantages when gamification is misused in education include: students thinking that they should only learn when they are going to receive some kind of external reward. Gamification is a multidisciplinary tool that encompasses various domains, theories of thought, methodologies, and various reasons for its implementation. Human beings are strongly attracted to games (Rocha *et. al.,* 2016).

Kahoot! is a digital application that can be used in the educational environment as a pedagogical proposal, enabling a lesson with different approaches and new assessment possibilities. Kahoot! allows students to learn using a technological device (computer or smartphone) while having fun. In this way, students play as if they were taking part in a game show, which allows them to learn in a fun, playful and interactive way, building their own knowledge (Kenski, 2012).

4.4.3. Materials and Methods

In this section we present the methodology used, which was a case study, with the instruments: direct observation of interactions and the students' reflections and perceptions. The research was applied to twenty-five students on the web programming course of the technical course in *internet* computing at IFG Câmpus Luziânia.

According to Yin (2003), case studies describe a phenomenon or intervention in the context in which it occurs. The literature review provided information on the subject and helped formulate the research questions. All the students reported that they had a smartphone and would use the institution's *internet* to carry out the activity. At the end of the activity, a *Google Docs* electronic questionnaire with open and closed questions was applied.

The form was filled in individually and included seven questions: What are the limitations of using mobile technology in the learning process? How did using the app during the activity contribute to your learning? What are the positive points of using mobile technologies in the learning process? What are the negative points of using mobile technologies in the learning process? What suggestions do you have for the activity? How satisfied are you with the methodology used in the activity? How satisfied are you with the application used in the activity?

The objectives of this research are exploratory and descriptive. Exploratory because it will allow the researcher to get closer to the problem, making it possible to refine ideas or discover new intuitions, and descriptive because some of the characteristics presented by the students will be exposed.

Content analysis was applied to the data collected from the open questions. Each answer was read more than once, coded and a frequency table was created. Themes were identified and, finally, the harmonisation of codes and themes was examined. Significant statements from the participants were included as quotes to illustrate. The closed answers were tabulated using Excel software and then analysed.

4.4.4. Results and Discussions

This section describes the results obtained in the survey, as shown in Table 31 and Figures 6 and 7.

Table 31. Questions asked of students and their answers.

Questions	Students' answers
What are the limitations of using mobile technology in the learning process?	Some may not have such advanced mobile phones or lack of battery power in their *smartphone* and are left out;
	The lack or speed of the *internet;*
	It can be very distracting, distracting from the real objective in the classroom;
	Technical faults;
	None;
How did using the app during the activity contribute to your learning?	Competition helps a lot;
	Using the mobile phone, they were able to carry out the activity and test their knowledge by interacting with the class;
	It made it easier to access the content I was studying;
	I think it's cool and creative;
	I liked it, it brought interaction with the class and debate on the content;
	It helps to be more attentive;
	It was fun, contributed dynamically and awakened the competitive side;
	It's more productive;
	It helped them to interact with their classmates and the teacher and to increase their interest;

	A quick quiz game helps you learn; It's fun and competitive, it makes you more interested in the subject; It helped with last month's revision; It made it easier for me to retain the content because it was an easy and fun way to revise;
	He contributed a lot;
What are the positive points of using mobile technologies in the learning process?	Learning becomes easier and more fun; Fun, practical, fast and with greater student interaction; An easier way to do research, communication, etc; If the student is interested, we agree 100%; Access to more content; It's done in a more relaxed way; Easier to learn, easier to communicate and study; It's a way of testing my knowledge; The ease and speed of accessing something; The dynamics and interest of the student are greater, the relationship between student and teacher is more interactive; They make lessons more intuitive; It increases communication, learning and didactic diversity; More familiarisation, accessible technology, fun and educational; Motivation to research the subject and excitement to attend class;
What are the negative aspects of using mobile technologies in the learning process?	I don't see any negative points; The case of using other apps or social networks that have nothing to do with the lesson and can distract some students; Technological faults, no good mobile phone, slow *internet,* low battery, sometimes it can crash and disturb the student; Sometimes we get more caught up in the competition
	than the content;

	Set time; The commitment to use the device for that purpose is sometimes not fulfilled, as it's easy to get distracted by other *software* on the mobile phone;
What suggestions do you have for the activity?	Do it more often with other subjects; It should be worth a point; The game couldn't use *internet* to load, as it crashes a lot; I'd like to see more of them, but over a longer period of time; Do more group activities that allow the use of mobile phones;

The students rated the Kahoot tool as interesting, mainly because of the game's intrinsic competitiveness. This demonstrates the importance of including games in teaching. Dellos (2015) corroborates this by saying that the competitiveness of the game makes the learning experience more valuable for the students. Also important is the positive impact it has had on interactions between the students, as well as between the teacher and the students. And all the students agreed that the time passed more quickly than in a traditional lesson. Dellos (2015) also states that Kahoot is a game that enables greater interaction between students, involving them and encouraging them not to give up easily when faced with difficulties in their subjects.

How satisfied are you with the methodology used in the activity?

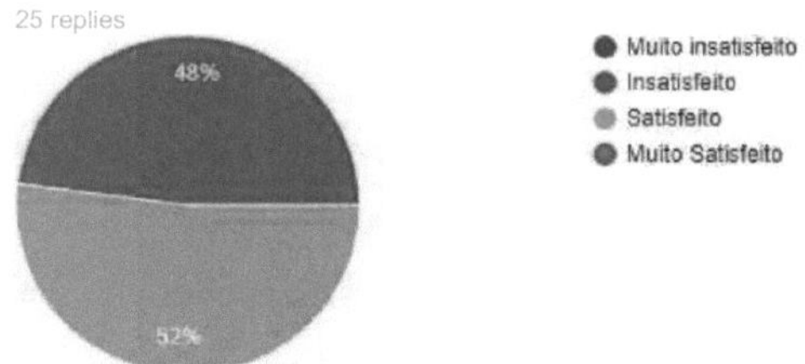

Figure 6. Level of satisfaction with the methodology used in the activity

Figure 6 shows that 48 per cent said they were very satisfied with the methodology used in the activity. The other 52% said they were satisfied with the methodology used in the activity. No student said they were dissatisfied or very dissatisfied with the activity.

How satisfied are you with the application used in the activity?

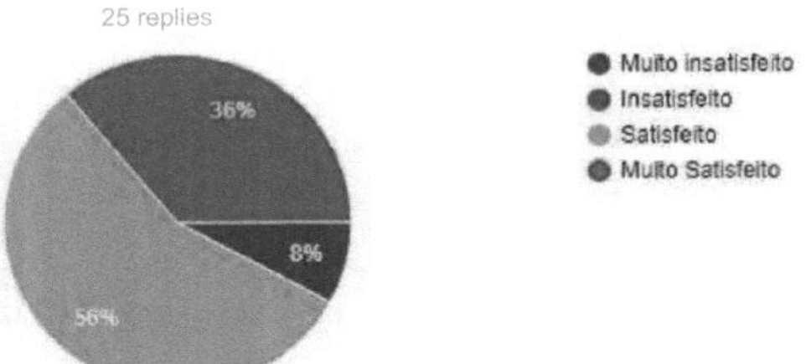

Figure 7. Level of satisfaction with the application used in the activity

Figure 7 shows that 56% said they were satisfied with the application used in the activity. 36% said they were very satisfied with the application used in the activity. And only 8% said they were very dissatisfied with the application used in the activity.

The teacher observed that the music and the score presented with each question made the game more stimulating. This confirms the study by Wang (2015) who assessed the concentration, enjoyment, engagement and fun of students playing Kahoot using or not using audio and scoring, and when audio and scoring were not used, the students did not maintain their focus and concentration in the same way. The presentation of the score on each question made the students stay more focused and engaged with the game, thus increasing competitiveness. Showing that music and points can have a significant influence on learning.

The game was easy to create and the website is practically self-explanatory. So there were no major difficulties in entering the questions. However, it was realised that having access to the *internet* is essential for using the tool. It was also noted that students can accidentally leave the game, and when they log in again, it is done as if they were a new user with the same name as the previous one and the score is not accumulated. Another limitation is the number of characters in the questions, making it necessary to write shorter questions.

During the game, the students remained very focused and apprehensive. They answered the questions quickly, and at the end of each question, they analysed how many had answered each alternative correctly and why the other answers were wrong, the groups that got it right cheered and everyone tried to become more competitive. When the groups had similar scores, they would try to answer the next question more quickly in order to overtake the other competing group.

This study confirms Wang (2015) who observed a significant increase in the engagement of students who used Kahoot in the classroom. In addition, most of the participating students considered that the software made lessons more fun and motivating.

4.4.5 Considerations

This case study reports on the experience of an activity developed on the Kahoot digital platform and analyses its potential for learning. It was clear from the responses that Kahoot had some limitations in terms of use, such as the availability of the *internet* in the classroom; however, Kahoot stimulated the students, making learning more attractive and fun. In this way, the students' perception and interest in the experience was positive, which places it at the level of a valid and recommended teaching strategy.

The findings of this study showed that the use of Kahoot in education affected students' attitudes, for example: it aroused enjoyment in class, generated new and positive learning experiences. Therefore, teachers can adopt this teaching strategy as a methodology that contributes to learning by making it fun, engaging, motivating and interesting. It can be said that most of the students enjoyed taking part in the activity.

CHAPTER 5

BIBLIOGRAPHICAL REFERENCES

Alencar, G. A.; Moura, M. R; Bitencourt, R. B. (2013). *Facebook* as a Teaching/Learning Platform: what IFSertão-PE teachers and students say. Education, Training & Technologies, v. 6, p. 86-93.

Alencar, G. A.; Pessoa, M. S.; Santos, A. K. F. S.; Carvalho, S. R. R.; Lima, H. A. B. (2015). *Whatsapp* as a teaching support tool. In: Congresso Brasileiro de Informática na Educação, p. 787795.

Amico, M. R. de A., Pra, R., Moraes, J. P. (2017). Applications of Kahoot! as an educational technology. In: 22nd Seminar on Education, Technology and Society, Taquara, RS. Electronic Proceedings of the Interdisciplinary Educational Journal (REDIN), p. 1-12.

Aragão, J. M. N; Gubert, F. do A.; Torres, R. A. M.; Silva, A. S. R. da; Vieira, N. F. C. (2018). The use of *Facebook* in health education: perceptions of adolescent students. Rev Bras Enferm. 71(2):286-292. DOI: http://dx.doi.org/10.1590/0034-7167-2016-0604.

Araújo, P. C; Bottentuit Junior, J. B (2015). The *WhatsApp* communication application as a strategy in Philosophy teaching, Temática (João Pessoa. On *line),* v. XI, p. 11-23.

Barbosa, M. F.; Barcelos, G. T.; Batista, S. C. F. (2015). "Inverted Classroom: Characterisation and Reflections". Information Technology Congress. vol. 8. N.p.

Barros, D. M. V. et. al. (2011). Education and technologies: reflection, innovation and practices. Lisbon: s.n.

Barseghian, T. (2011). "Three Trends That Define the Future of Teaching and Learning", https://www.kqed.org/mindshift/50640/how-reading-novels-in-math- class-can-strengthen-student-engagement (30 March 2018).

Batchelor, J.; Herselman, M.; Traxler, J. (2010). Teaching and learning with new technology: a tough nut to crack. IST-Africa 2010, vol. 1, no. 7, pp.1-7, 19-21 May.

Bedi, K. (2014). Tablet PC & smartphone uses in education (Tablet Tours). In: 37th International Convention on Information and Communication Technology, Electronics and Microelectronics (MIPRO). p. 940-945.

Beland, L.; Murphy, R. (2015). III Communication: Technology, Distraction & Student Performance.

Bergmann, J.; Sams, A. (2016). A. Flipped classroom: an active learning methodology. 1 ed. Rio de Janeiro: LTC.

Bergmann, J.; Sams, A. (2012). Flip Your Classroom: Reach Every Student in

Every Class Every Day, USA, International Society for Technology in Education.

Bishop, J. L. (2013). A Controlled study of the flipped classroom with numerical methods for engineers. 284 f. Thesis (Doctorate in Engineering Education) - UTAH State University, Logan.

Bishop, J. L.; Verleger, M. A. (2013). The Flipped Classroom: A Survey of the Research. In: ASEE Annual Conference & Exposition, 120, Atlanta. Proceedings... Washington DC, American Society for Engineering Education.

Boll, C. I.; Lopes, R. C.; Luchini, N. A. (2016). Mobile technologies and distance education: more than creating apps, you need to know what to do with them. International Symposium on Distance Education and Meeting of Researchers in Distance Education - SIED: ENPED:2016, p. 01-11.

Borges, P. F. B. (2015). New Digital Information and Communication Technologies
Applied to Secondary and Technical Education at a Federal Public School in Uberaba - MG. 2015. 158 f. Dissertation (Master's in Education) - Federal University of the Triângulo Mineiro, Uberaba (MG).

Brown, R. B. (2006). Doing your Dissertation in Business and Management: The Reality of Researching and Writing. SAGE, London.

Burke, B. (2015). Gamify: how gamification motivates people to do extraordinary things. Translation by Sieben Grupe. DVS Publishing. São Paulo.

Cabral, T. C. B. (2005). Teaching and Learning Mathematics in Engineering and the Use of Technology. CINTED-UFRGS, Rio Grande do Sul, v. 3, n. 2, p. (unpaginated).

Castells, M. (2003). The *Internet* galaxy: reflections on the *Internet,* business and society. Rio de Janeiro: Zahar.

Coil, D. A.; Ettinger, C. L.; Eisen, J. A. (2017). Gut Check: The evolution of an educational board game. PLOS Biology, 15(4), e2001984.

Brazilian *Internet* Steering Committee (2014). ICT Household Survey 2013: research on the use of information and communication technologies in Brazil. São Paulo: Brazilian *Internet Steering Committee* [Internet] (04 June 2018). http://www.cetic.br/media/docs/publicacoes/2/TIC_DOM_EMP_201 3 _electronic_book.pdf

Cordeiro, S. F. N.; Bonilla, M. H, S. (2015). Mobile digital technologies: reterritorialisation of everyday school life. Educar em Revista, Curitiba, n. 56, p. 259-275.

Coscarelli, C. V.; Ribeiro, A. E. (eds.) (2011). Digital literacy: social aspects and pedagogical possibilities. 3ª ed. - Belo Horizonte: Ceale; Autêntica Editora.

Costa, E.; Silva, A. P.; Cordeiro, B. M. P.; Silva, C. A. (2014). Digital technologies have arrived! What to do? Innovative ways of learning. In: DANTAS, L. G.; MACHADO, M. J. (Org.).

Technologies and education: perspectives for management, knowledge and teaching practice. 2 ed.

Dellos, R. (2015). Kahoot! A digital game resource for learning. International Journal of Instructional Technology and Distance Learning, 12(4), 49-52.

Deterding, S.; Dixon, D.; Khaled, R.; Nacke, L. (2011). From game design elements to gamefulness: defining gamification. In: Proceedings of the 15th International Academic MindTrek Conference - MindTrek'11, p. 9-11.

Dicheva, D.; Dichev, C.; Agre, G.; Angelova, G. (2015). Gamification in Education: A Systematic Mapping Study. Educational Technology & Society. v. 18, n. 3, p. 1-14.

Domínguez, A.; Saenz-de-Navarrete, J.; de-Marcos, L.; Fernández- Sanz, L.; Pagés, C.; Martínez-Herráiz, J.J. (2013). Gamifying learning experiences: Practical implications and outcomes.
Computers & Education. 63. pp. 380- 392

Facebook (2018). Principles. (04 June 2018). https://www.facebook.com/principles.php.

Faria, V. P; Costa, H.; Parreira Junior, P. A. (2016). eQuest: A Gamified Student Response System. In: Brazilian Congress of Informatics in Education. p. 280-287.

Feenberg, A. (2010). The factory or the city: which model of distance education via the web? In: NEDER, Ricardo T. Andrew Feenberg's critical theory: democratic rationalisation, power and technology.

Ricardo T. Neder (org.) Brasília: Observatory of the Movement for Social Technology in Latin America/CDS/UnB/Capes, p. 153-175.

Feitosa, T.; Machado, L. (2014). School institutes zero tolerance for mobile phones in the classroom. Gazeta On *line.* 26 Aug.

Ferreira, J. L; Correa, B. R. P. G; Torres, P. L. (2012). The pedagogical use of the social network *Facebook.* Colabor@, v. 7, p. 1-16. (04 June 2018). http://pead.ucpel.tche.br/revistas/index.php/colabora/ article/view/199/152.

Figueiredo, M.; Paz, T.; Junqueira, E. S. (2015). Gamification and education: a state of the art of research carried out in Brazil. In: Congresso Brasileiro de Informática na Educação. p. 1154-1163.

Frescki, F. B.; Pigatto P. (2009). Difficulties in learning Differential and Integral Calculus in Technological Education: a proposal for a Levelling Course. I Simpósio Nacional de Ensino de Ciência e Tecnologia, Ponta Grossa, p.910-917.

Gadotti, M. (2000). Current perspectives on education. São Paulo em Perspectiva, v. 14, n. 2, p. 3-11, 2000.

Gannod, G.C.; Burge, J.E.; Helmick, M. T. (2008). Using the inverted classroom to teach software engineering. In: ACM/IEEE 30th International Conference on

Software Engineering, 2008. Leipzig: Proceedings of the 30th international conference on Software engineering. p. 777-786.

Gazotti-Vallim M. A. (2017). Experiencing English with Kahoot. The ESPecialist: Description, Teaching and Learning, 38(1):1-18.

Gil, A. C. (1999). Methods and techniques of social research. São Paulo: Atlas.

Gil, A. C. (2002). How to design research projects. 4. ed. São Paulo: Atlas.

Gómez, A. (2015). Perez. Education in the digital age - the educational school. Porto Alegre: Penso.

Gonçalves, H. A.; Nascimento, M. B. C.; Nascimento, K. C. S. (2015). Systematic Review and Meta-analysis: levels of evidence and scientific validity. Revista Eletrónica Debates em Educação Científica e Tecnológica, v. 5, p. 193-211.

Guenaga, M. et. *al.* (2012). *Smartphones* and teenagers, threat or opportunity. In: 15th International Conference on Interactive Collaborative Learning, IEEE, p. 1-5.

Honorato, W. A. M.; Reis, R. S. F. (2014). *WhatsApp* - a new tool for teaching. In: Proceedings of the IV Symposium on Development, Technologies and Society, p. 1-6.

Jaime, M. P.; Koller, M. R. T.; Graeml, F. R. (2015). The application of flipped classroom in the strategic management course. In: Jornadas Internacionales de Innovación Universitaria Educar para Transformar, 12, 2015. Proceedings... Madrid: Universidad Europea, p. 119-133.

Juliani, D. P.; Juliani, J. P.; Souza, J. A. de; Bettio, R. W. (2012). Using social networks in education: a guide to using *Facebook* in a higher education institution. RENOTE. Revista Novas Tecnologias na Educação, v. 10, p. I-XI.

Kapp, K. M. (2012). The gamification of learning and instruction: game-based methods and strategies for training and education. San Francisco: Pfeiffer.

Kenski, V. M. (2012). Education and technologies: the new pace of information. 8. ed.
Campinas, SP: Papirus.

Kitchenham, B. (2004). Procedures for performing systematic reviews. Technical Report TR/SE-0401, Keele University and NICTA.

Knobel, M.; Lankshear, C. A. (2007).
New Literacies Sampler. New York: Peter Lang.

Lage, M. J.; Platt, G. J.; Treglia, M. (2000). Inverting the classroom: a gateway to creating an inclusive learning environment. Journal of Economic Education. Bloomington, IN, v. 31, n. 1, p. 30-43.

Lee, J. J.; Hammer, J. (2011). Gamification in Education: What, How, Why Bother? Academic Exchange Quarterly. v. 15, p. 1-5.

Lee, V.; Schneider, H.; Schell, R. (2004). Mobile applications: architecture, design, and development. Prentice Hall.

Lemos, A.; Josgrilberg, F. (2009).Communication and Mobility. Sociocultural Aspects
of Mobile Technologies in Brazil. Salvador: Edufba.

Lévy, P. (2010). *Cyberculture.* Translation by Carlos Irineu da Costa - São Paulo: Editora 34 Ltda.

Lima, J. O.; Andrade, M. N.; Damasceno, R. J. A. (2010). Teacher resistance in the face of new technologies, p. 1-4. (24 May 2018) http://meuartigo.brasilescola.com/educacao/a-resistencia- professor-diante-das-novastecnologias.htm.

Lima, J. R.; Capitão, Z. (2003). E-learning and e-content. Lisbon: Atlantic Centre.

Llorens, F.; Capdeferro, N. (2011). Posibilidades de la plataforma *Facebook* para el aprendizaje colaborativo en línea [artículo en línea]. University and Knowledge Society Journal (RUSC). Vol. 8, No. 2, pp. 31-45. UOC.

Lopes, C. G.; Vaz, B. B. (2016). The Pedagogical Use of *Whatsapp* Groups in History Teaching. In: 5th International Congress of History - New Epistemes and Contemporary Narratives, p. 1-28, Jatai. 5th International History Congress - New Epistemes and Contemporary Narratives.

Lucena, S. (2016). Digital cultures and mobile technologies in education. Educar em Revista, Curitiba, Brazil, n. 59, p. 277-290.

Marçal, E.; Andrade, R.; Rios, R. (2005). Learning using mobile devices with virtual reality systems. CINTED- UFRGS, v. 3, n. 1, Porto Alegre: May.

Mattar, J. (2012). The use of networks in education. (03 June 2018). http: //www. educacaoetecnologia. org.br/?p=5487.

Mattar, J. (2013). Web 2.0 and social networks in education. São Paulo: Artesanato educacional.

Merije, W. (2012). Mobimento: mobile education and communication. São Paulo: Peirópolis.

Miranda, L. A. V. (2005). Online education: interactions and learning styles of higher education students on a web platform. 2005. 382 f. Thesis (Doctorate in Education) - University of Minho, Braga.

Moran, J. (1995). New Technologies and the Reenchantment of the World. Educational Technology Magazine. Brazil, vol. 23, n. 126, p. 24-26.

Moran, J. M. (2013). New technologies and pedagogical mediation. 21 ed. rev. and current. - Campinas, SP: Papirus.

Moran, J. M. (2014). New personality [25 October 2014]. Brasília: Correio Braziliense. Brasília. Interview granted to Olivia Meireles. http://www2. eca.usp.br/moran/wp- content/uploads/2014/01/Jos%C3%A9-Moran.pdf. (20 June 2018).

Moran, J. M. (2015). Hybrid education: A key concept for education today. In: Bacich, Lilian; Neto, Adolfo Tanzi; Trevisani,

Fernando de Mello. Hybrid teaching: personalisation and technology in education. Porto Alegre: Penso.

Moran, K.; Milsom, A. (2015). The Flipped Classroom in Counselor Education. Counsellor Education and Supervision.

Moreira, A. F. B.; Kramer, S. (2007). Contemporaneity, Education and Technology. Educ. Soc., Campinas, vol. 28, n. 100 - Especial, p. 1037- 1057. http://www.scielo.br/pdf/es/v28n100/a1928100.pdf (12 April 2018).

Morrison, B. B.; Disalvo, B. (2014). Khan academy gamifies computer science. In: Proceedings of the 45th ACM Technical Symposium on Computer Science Education - SIGCSE '14, p. 39- 44.

Moura, A. (2009). Geração Móvel: um ambiente de aprendizagem suportado por tecnologias móveis para a "Geração Polegar", Emerging Environments, VI International Conference on ICT in Education, p. 49-77.

Mulbert, A. L.; Pereira, A. T. C. (2011). An overview of research into mobile learning *(m-learning).* In: Associação Brasileira de Pesquisadores em *Cibercultura,* Florianópolis.V Simpósio Nacional da ABCiber.

Niza, C. (2016). How to use *WhatsApp* at school. Blog Tecnologia na Educação. https://novaescola.org.br/conteudo/4688/como-usar-o- whatsapp-na-escola. (17 April 2018).

Nyiri, K. (2002). Towards a philosophy of *m-Learning.* In: IEEE International Workshop On Wireless And Mobile Technologies In Education - WMTE.

Oliveira, E. D. S.; Medeiros, H.; Leite, J. E. R.; Anjos, E. G.; Oliveira, F. S. (2014). Proposal for a model of courses based on *mobile learning:* an experiment with teachers and tutors on *WhatsApp,* p. 3482-3496. In: Proceedings of the XI Brazilian Congress of Distance Higher Education.

Padrón, C. J. (2014). Didactic Strategies based on *WHATSAPP* Instant Messaging Applications exclusively for Mobiles: (Mobile *Learning)* and the use of the Tool to promote Collaborative *Learning.* Eduweb, v. 7, n. 2, p. 123134.

Paiva, L. F. De; Ferreira, A. C.; Corlett, E. F. (2016). The use of *WhatsApp* as a didactic-pedagogical communication tool in higher education. In: Workshops of the Brazilian Congress of Informatics in Education, p. 751-760.

Park, S.; Cho, K.; Lee, B. G. (2014). What Makes Smartphone Users Satisfied with the Mobile Instant *Messenger?:* Social Presence, Flow, and Self-disclosure. International Journal of Multimedia & Ubiquitous Engineering, v. 9, n. 11.

Pavanelo, E.; Lima, R. (2017). Inverted classroom: analysing an experience in Calculus I. Bolema, Rio Claro, v. 31, n. 58, p.739-759.

Pereira, L. R; Schuhmacher, V. R. N; Schuhmacher, E; Dalfovo, O. (2012). The use of technology in education, prioritising mobile technology (03 June 2018). http://www. senept.cefetmg.br/galerias/Anais_2012/GT-02/GT02- 014.pdf

Pereira, P. C.; Pereira, R. S.; Alves, J. C. (2015). Virtual environments and communication media, addressing the explosion of media in the information society and its impact on learning - the use of *WhatsApp* as an *m-learning* platform. Revista Mosaico. Jan./Jun.; 06 (1): 29-41.

Pereira, S. R. C.; Pimentel, E. P. (2014). Gamified Virtual Laboratory for Teaching Chemistry on Mobile Devices. In: Workshop on Mobile Technologies in Education. p. 396-405.

Pompeo, C. (2014). Teachers vie for students' attention with social networks. Gazeta do Povo. Londrina, 24 May.

Porvir. (2013). 10 tips and 13 reasons to use mobile phones in class. Porvir website. http://porvir.org/porfazer/10-dicas-13-motivos-para-usar- celularna-aula/20130225. (12 April 2018).

Prensky, M. (2012). Learning based on digital games. São Paulo: Senac.

Ramal, A. (2015). Inverted classroom: the education of the future. Rio de Janeiro: G1 Educação, http://g1.globo.com/educacao/blog/andrea- ramal/post/sala-de-aula-invertidaeducacao-do-futuro.html (25 June 2018).

Rambe, P.; Bere, A. (2013). Sing mobile instant messaging to leverage learner participation and transform pedagogy at a South African University of Technology. British Journal of Educational Technology. 44, 4, 544-561, July.

Rocha, P. S. R.; Lima, R. W. de; Macedo, R. L.; Leite, C. R. M. Neto, F. M. M. (2016). Gamification: An application for teaching Brazilian Sign Language. In: V Congresso Brasileiro de Informática na Educação - CBIE 2016, p. 896-900.

Royle, K.; Stager, S.; Traxler, J. (2014). Teacher development with mobiles: Comparative critical factors. Prospects, v. 44, n. 1, p. 29-42.

Saboia, J.; Vargas, P. L.; Viva, M. A. A. (2013). The use of mobile devices in the teaching and learning process in the virtual environment. Revista Cesuca Virtual: Conhecimento Sem Fronteiras v.1, n. 1. p. 1-13. (03 June 2018). http://ojs.cesuca.edu.br/index.php/cesucavirtual/article/view/424/209 .

Saccol, J. B. A.; Schlemmer, E.; Barbosa, J. (2011). *M-learning* and u-*learning*: new perspectives on mobile and ubiquitous learning. São Paulo: Pearson Prentice Hall.

Sampaio, R. F.; Mancini, M. C. (2007). Systematic review studies: a guide for careful synthesis of scientific evidence. Revista Brasileira de Fisioterapia, v. 11, n. 1, 83-89.

Santa Catarina (2008). Law no. 14.363, of 25 January 2008. Prohibits the use of mobile phones in state schools in the state of Santa Catarina. Santa Catarina: Florianópolis, 25 Mar.

Santana, R. C. M. *et. al.* (2016). The use of mobile technologies in science teaching: an experience on the study of coastal ecosystems in the Atlantic Forest of southern Espírito Santo. Revista Ibero-Americana de Estudos em Educação, Araraquara, v. 11, n. 4 p. 2234-2244.

Santi, A. de; Garattoni, B. (2015). The dark side of *Facebook.* Super Interessante magazine (03 June 2018). http: //super. abril .com.br/superarquivo/348.

Schneider, E. I., Suhr, I. R. F., Rolon, V. E. K. E. Almeida, C. M. de (2013). Inverted Classroom in Distance Learning: a Blended Learning proposal. In Revista Intersaberes, n. 16, v. 8, p. 68-81.

Seabra, C. (2010). Technologies at school. Porto Alegre: Telos Empreendimentos Culturais.

Sena, D; Burgos, T. (2010). The computer and the mobile phone in the teaching-learning process of school physical education. In: 3rd Hypertext and Technologies in Education, Pernambuco. Proceedings of the Hypertext Symposium.

Sharples, M.; Taylor, J.; Vavoula, G. (2007). A Theory of Learning for the Mobile Age. In R. Andrews and C. Haythornthwaite (eds.) The Sage Handbook of Elearning Research. London: Sage, pp.221-247.

Silva, M. (2012). Interactive classroom: education, communication, classic media... 6.ed. São Paulo: Edições Loyola.

Singh, K. (2007). Quantitative Social Research Methods. SAGE Publications, New Delhi.

Souza, C. F. de. (2015). Distance learning: mobile digital technology in English language teaching. Revista Texto Livre, v. 8, p. 3950.

Staker H.; Horn M. B. (2013). Classifying K-12 Blended Learning. http://files.eric.ed.gov/fulltext/ED535180.pdf. (30 April 2018).

Tapscott, D. (2010). The time of the digital generation: how young people who grew up using the *internet* are changing everything, from companies to governments. Trad. Marcello Lino. Rio de Janeiro: Agir Negócios.

Tardif, M. (2011). Teaching knowledge and professional training (12. ed.). Petrópolis, RJ: Vozes.

Tarnopolsky, O. (2012). Constructivist blended learning approach to teaching english for specific purposes. Berlin: De Gruyter Open.

Teixeira, G. P. (2013). Flipped classroom: a contribution to the learning of Camonian lyric. 2013. 167 f. Dissertation (Master's in ELearning Systems Management) - Faculty of Social Sciences and Humanities, Universidade Nova Lisboa, Lisbon.

Thiollent, M. (2011). Action research methodology. 18.ed. São Paulo: Cortez.

Trevelin, A. T. C.; Pereira, M. A. A and Oliveira Neto, J. D. de. (2013). The Use of the 'Inverted Classroom' in Higher Technology Courses: Comparison Between the Traditional Model and the Inverted 'Flipped Classroom' Model Adapted to Learning Styles". In Revista de Estilos de Aprendizagem, n. 12, v. 11, p. 1-14.

Tune, J. D; Sturek, M.; Basile, D. P. (2013). Flipped classroom model improves graduate student performance in cardiovascular, respiratory, and renal physiology. Adv Physiol Educ, Indianapolis, v. 37, n. 4, p. 316-320.

Unesco (2014). Policy guidelines for mobile learning. United Nations Educational, Scientific and Cultural Organisation (UNESCO), France.

Unicef (2013). United Nations Children's Fund. *Internet* use by adolescents. Brasília, DF: UNICEF. (04 June 2018) http://www.unicef.org/brazil/pt/br_uso_internet_adolescentes.pdf.

Valente, J. A. (2012). The Intelligent Use of Computers in Education. NIED, UNICAMP.

Valente, J. A. (2014). Blended Learning and Changes in Higher Education: the Inverted Classroom Proposal. In Educar em Revista, Curitiba, PR, Special Edition, n. 4, p. 79-97, Editora UFPR.

Vavoula, G. N., & Sharples, M. (2002). KLeOS: A personal, mobile, Knowledge and Learning Organisation System. In Milrad, M., Hoppe, U. Kinshuk (eds.) Proceedings of the IEEE International Workshop on Mobile and Wireless Technologies in Education (WMTE2002), Aug 29-30, Vaxjo, Sweden, pp. 152- 156.

Wains, S. I.; Mahmood, W. (2008). Integrating *m-learning* with e- learning. 9th ACM SIGITE Conference on Information Technology Education, Cincinnati, USA, pp. 31-38.

Wang, A. I. (2015). The wear out effect of a game-based student response system. Computers and education. vol. 82.

Yin, R. K. (2003). Case Study Research: Design and Methods (3rd Ed.). Thousand Oaks, CA: Sage Publications. 181p.

MIX
Papier aus verantwortungsvollen Quellen
Paper from responsible sources
FSC® C105338

Printed by Books on Demand GmbH, Norderstedt / Germany